T0323651

Sustainable Governance
of Natural Resources

Sustainable Governance of Natural Resources

*Uncovering Success Patterns
with Machine Learning*

ULRICH FREY

OXFORD
UNIVERSITY PRESS

OXFORD
UNIVERSITY PRESS

Oxford University Press is a department of the University of Oxford. It furthers the University's objective of excellence in research, scholarship, and education by publishing worldwide. Oxford is a registered trade mark of Oxford University Press in the UK and certain other countries.

Published in the United States of America by Oxford University Press
198 Madison Avenue, New York, NY 10016, United States of America.

Library of Congress Cataloging-in-Publication Data
Names: Frey, Ulrich, 1975– author.
Title: Sustainable Governance of Natural Resources :
Uncovering Success Patterns with Machine Learning / Ulrich Frey.
Description: New York, NY : Oxford University Press, [2020] |
Includes bibliographical references and index.
Identifiers: LCCN 2020033439 | ISBN 9780197502211 (hardback) |
ISBN 9780197502235 (epub) | ISBN 9780197502228 (UPDF) |
ISBN 9780197502242 (Digital online)
Subjects: LCSH: Sustainable development. | Natural resources—Management.
Classification: LCC HC79.E5 F7328 2020 | DDC 333.70285/631—dc23
LC record available at https://lccn.loc.gov/2020033439

1 3 5 7 9 8 6 4 2

Printed by Integrated Books International, United States of America

To my parents

Contents

Acknowledgments

I would like to express my gratitude to Claudia Pahl-Wostl, who had the idea to make my work available to the English-speaking community.

I also would like to thank Lin Ostrom for inspiring this kind of research - without her very generous support this book would not have been written!

A big thanks goes to Sonja, Johannes, Anne, Felix, Elinor, and Laura for their help! I also wish to thank Joachim, Hans, and Maria for their support.

1

Introduction

1.1 The high importance of natural resources

We know that the management of natural resources is very important, not the least because many resources are finite and may eventually run out (Meadows et al. 1972). This makes their sustainable management mandatory (Brundtland et al. 1987). We will always be confronted by two questions: (1) How can more and more people be fed from a land area that remains constant? (2) How can natural resources such as fish stocks or forests be managed in a way that future generations may harvest from them as well? These two questions have gained more and more relevance, and as a result, sustainability is on the political agenda on all levels—local, national (Bundesregierung 2015), and global (e.g., the seventeen sustainable development goals, see Biermann, Kanie, and Kim 2017). However, sustainable management of natural resources is often seen as the opposite of growth, decreasing economic benefits. Today, we understand that a continuous growth, for example, in expanding fisheries will result in overfished fishing grounds and to a decrease in stocks.

Consequently, it is high time to understand how we can determine possible success patterns for natural resources and manage such resources sustainably. While some resources may be replaced by other products, this is hardly conceivable for fish or other sorts of food. Hence, this study concentrates on natural resources which are exhaustible by extraction, that is, biotic resources such as fish or forests (Baland and Platteau 1996). Second, we concentrate on resources on a local scale. A primary goal of this investigation is to develop a synthesis of possible success patterns for natural resources based on a new theoretical, empirical, and methodological approach.

1.1.1 Natural resources in danger of depletion

The concept of sustainable management has been understood quite differently across disciplines. There are many different sustainability concepts, and many different aspects have been subordinated to these concepts (Kopfmüller 2001). The present analysis understands sustainability very narrowly: as the sustainable,

Sustainable Governance of Natural Resources. Ulrich Frey, Oxford University Press (2020). © Oxford University Press.
DOI: 10.1093/oso/9780197502211.001.0001.

that is, continuing, use of the three natural resources studied without depleting them: fish, forests, and water.

Thus, we focus on the ecological dimension for a higher analytical sharpness. Some other dimensions—such as participation in social decision-making processes or fair distribution of environmental use opportunities—are included in the analyses as prerequisites for ecological success only, as is the case in various other models (Kopfmüller 2001).

In recent years and decades, the pace at which natural resources have been exploited or even destroyed has increased dramatically. This is true despite increased protection efforts (FAO 2015) and humankind's growing awareness of the problem (Taylor 1986). This is mostly due to an increasing need of human space and associated use (Sala et al. 2005). For all three resource types studied in this paper—fish, forests, and irrigated land—the situation has now become dramatic.

For example, for fisheries (see Figure 1.1), across fish species and habitats, complete overfishing in all oceans is a reality (Myers and Worm 2003).

The data in Figure 1.1 are more than alarming, proving that entire ecosystems are in danger of being depleted. Due to industrial fishing, in only about fifteen years we could witness the worldwide decrease of biomass of large fish predators by around 80% (He et al. 2019; Myers and Worm 2003). Moreover, possible future developments are seen very pessimistically: it is highly likely that habitat loss and high biodiversity losses will occur (Sala et al. 2005). Yet, the importance of biodiversity for the functioning of ecosystems is enormous and is still strongly underestimated (MacDougall et al. 2013; Soliveres et al. 2016).

Unfortunately, the global downtrend for forests does not show a brighter picture: Since the beginning of settlement, mankind has cut down more than 50% (30 million km^2) of existing forests (Sunderlin et al. 2005) and this process is speeding up alarmingly. The gross forest cover loss for a period of only five years (2000–2005) is 3.1% (Hansen, Stehman, and Potapov 2010). This corresponds to more than 1 million km^2 of forest area, about three times the area of the Germany. For some decades now, the loss of forest area has been around 50,000–110,000 km^2 per year in tropical forests (Achard et al. 2014). Recent developments in the Amazon region also give cause for concern—deforestation continues unabated, since between 2010 and 2015 320,000 km^2 have been cut down in tropical forests (Diaz Sandra et al. 2019).

Increasing problems also exist in the third resource system studied, irrigation systems. These systems use a large part (90%) of global freshwater resources. With a growing world population, the agricultural production grows as well, which leads inevitably to an increase in irrigated land and an increase in water consumption both for agricultural and personal use. Consequently, global water supply bottlenecks (water stress) do occur. In some regions, water stress

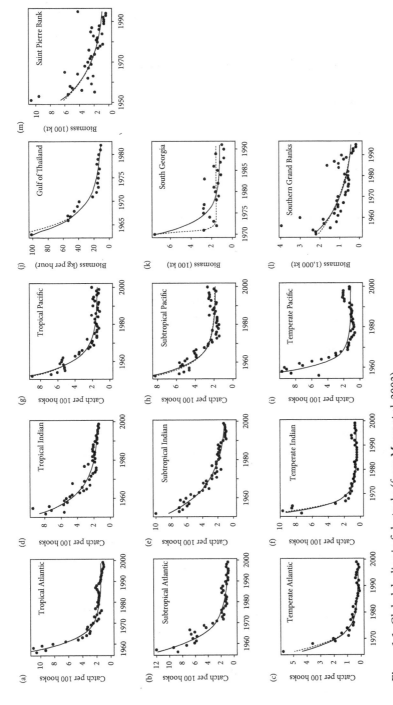

Figure 1.1 Global decline in fish stocks (from Myers et al. 2003)

is additionally influenced by climate change (Alcamo, Flörke, and Märker 2007; Vörösmarty et al. 2000; Black and King 2009; Tang 1992).

This loss of natural resources goes hand in hand with a loss of genetic diversity. Biodiversity loss is noticeable at different trophic levels (Dobson et al. 2006). Dramatic losses have been reported for insects, with more than 75% lost worldwide (Hallmann et al. 2017; Seibold et al. 2019). Hence, biodiversity loss is one important indicator for (un)sustainable management. In this analysis, we will use it as an indicator for forest condition, that is, as a proxy for the successful management of forests.

To sum up: in the year 2020, it is a well-established fact that natural resources deteriorate world-wide (IPCC 2018). Globally, three developments put the sustainable use of many natural resources in great danger: climate change, an increasing population, and our expanding land consumption. These processes are primarily anthropogenic in origin and irreversible at a certain degree of change. This irreversibility greatly reduces the robustness of ecological systems (resilience). Thus, it can be shown, for example, that there are three stable states (basins of attraction) for tree cover: forest (>60% tree cover), savannah (<60% and >5%), and no trees at all (>5%). If human-made climate change results in lower rainfall and deforestation leads to less than 60% of the forest cover, a self-reinforcing process of transition from forest to savannah is triggered (Hirota et al. 2011).

This book is about management of natural resources by local communities. This form of management is very important globally: Fisheries support two hundred million livelihoods (Cinner et al. 2012), 18% of all forests are community managed (White and Martin 2002), and irrigation systems using up to 90% of global freshwater consumption produce 40% of global food production (Siebert et al. 2010). Therefore, it is extremely relevant for sustainable management that we know what causes these systems to fail or succeed.

1.1.2 Social-ecological systems

The management of natural resources takes place at the interface of ecological systems, such as fish stocks, and their social rule systems, such as catch quotas. These systems are called social-ecological systems (SES, Ostrom 1990) and include many different sectors of natural resources such as forests or fisheries on various geographical scales. Since technical infrastructure often play a role as well, it is possible to speak of SETS (social-ecological-technical systems).

The present investigation is limited to local resources, that is, it does not go beyond a regional scale. We use the following, frequent definition: "A SES is an ecological system intricately linked with and affected by one or more social systems" (Anderies, Janssen, and Ostrom 2004, p. 20).

This view of natural resource management attaches particular importance to the interaction between biophysical attributes, social variables, and institutional structures. Precisely because it has been shown that a one-sided view is not enough to do the complexity of these systems justice, the present analysis is orientated toward such an integrated and interdisciplinary approach (Ostrom 1990; Pahl-Wostl et al. 2013; Meinzen-Dick 2007).

Ignoring one side of SES has led to repeated failures in the past. Particularly striking are one-sided, top-down approaches to development cooperation. Thus, for example, a few decades ago, the government agency in Nepal (department of irrigation, DOI), which is responsible for irrigation systems, considered the improvement of technical infrastructure as the most important factor. Social factors were hardly considered or taken into account. Although the infrastructure of many irrigation systems has improved enormously with tremendous capital investments, results have lagged far behind expectations (Lam 1998; Tang 1992).

Previous government attempts to improve resource management in such SES often failed to reach the community level since the importance of social and institutional systems was not always recognized and traditional social structures and rule systems were not taken into account. Even if infrastructure was improved and management responsibilities were handed over to local farmers, these measures often proved to be inadequate. If neither appropriate local rules nor organizational structures existed, external action did not meet with success (Lam 1998; Tang 1992). Hence, despite many efforts of national governments to establish uniform policies and procedures for agriculture, forest management, and fisheries, many users and resource systems are not affected because of their small size and geographical location (Altieri 2002).

Thus, SES research stresses the importance of paying attention to the role of the local community and their measures. Moreover, we are warned not to try to improve the success of such systems through technological improvements alone. Improved headworks in irrigation systems will not automatically improve results. The research strand of community-based natural resource management focuses on this insight (Blythe et al. 2017; Brooks, Waylen, and Borgerhoff Mulder 2012; van Laerhoven 2010). Building on these experiences, this research field focuses on the organization of communities, taking into account the interactions between technologies, social attributes, and rule systems (Gruber 2008; Pagdee, Kim, and Daugherty 2006). If the role of learning and the adaptation of strategies is particularly emphasized, it is called adaptive co-management (Plummer and Armitage 2007). I'll come back to it later (see Section 2.4.3.2).

Due to these interactions of system attributes SES is *complex* (Limburg et al. 2002), often a major obstacle in their analysis (cf. Folke 2006 for different dimensions of complexity). Since many system attributes interact with one another, often enough not in a linear manner, the creation of a standard model for ecological success appears to be problematic particularly because causal relationships between system attributes have so far been neglected (Agrawal 2001). The neural networks model used in this analysis takes these problems into account (see Section 4.1.3).

There are various forms managing complex SES (see Section 2.4.1). An important form is the management as common property. This is called common pool resource (CPR) management. It is very common (about 18% of forest area in forests) and of great importance for the subsistence of many people worldwide (FAO 2015; Agrawal 2007; White and Martin 2002). This form of organization of natural resources as common property requires cooperation to a special degree and is dealt with separately in Section 2.3. It often arises when the yield of a resource is limited or when the group's joint efforts are needed to use the resource efficiently (Ostrom 1990). In practice, one is interested in establishing cooperation as stable and as long-standing as possible in order to achieve all associated advantages such as higher economic efficiency, ecological sustainability, and social justice.

1.2 Research question and goals

1.2.1 Goals and benefits

SES are usually studied via individual case studies, that is, a resource system is analyzed in detail by collecting data—mostly via interviews, but also with many other methods. Many of the variables collected are usually only applicable to the system under investigation. Thus, existing detailed knowledge about many thousands of systems is very high (Poteete, Janssen, and Ostrom 2010). On the other hand, comparability suffers from this small-scale approach, since the methodology used and the variables collected differ between studies.

This problem is aggravated since many examples of very similar systems show little similarity regarding their success. Structural similarity here refers to factors such as the same size of the system, a comparable history, or the same social and political framework. Accordingly, a particular theoretical challenge is to explain the differences between failures and success stories of local communities which have given themselves rule systems in the sense of sustainability, efficiency, and justice.

Despite some meta-analyses (Pagdee, Kim, and Daugherty 2006; Gutiérrez, Hilborn, and Defeo 2011; Brooks, Waylen, and Borgerhoff Mulder 2012), it is still largely open whether there is a set of factors for SES that can explain either ecological success or failure in the management of natural resources. The aim of the present analysis is to quantify empirically those factors that make SES ecologically successful.

It has often been doubted whether individual case studies could be generalized (e.g., Cleaver 2000). In this book we try to create general models for complex SES with satisfactory explanatory quality. If this were not the case, the pessimistic stance would be justified—a generalization of case studies beyond the local context would not be possible.

In contrast, if models with a large data base reach a satisfactory predictive accuracy, the hypothesis would be strengthened that there are general success patterns in natural resource management. Therefore, the aim of the present analysis is to create quantitative models which predict as precisely as possible the ecological success in the use of natural resources. For this goal we use a comprehensive set of factors.

This requires at least three prerequisites: first, a broad data base of case studies that have a common set of variables for comparisons, both comparable and comprehensive. Second, a conceptualization of possible success factors on an abstract, cross-case study level. Third, a methodology that is able to handle complex, non-linear data with an unknown model structure.

To fulfill these requirements, it is first necessary to develop a theoretical model from existing theoretical studies and data (this is implemented in Section 5.1). Such a comprehensive conceptualization or framework does not exist yet (but see Ostrom 2009). On the one hand, this conceptualization must be abstract enough to be used generally, that is, across sectors. On the other hand, it must be concrete enough so that empirical data can be assigned unambiguously to individual categories within the framework. Hence, variables, indicators, and categories each have to be arranged according to their correct level of abstraction.

Once models with a satisfactory explanatory quality exist based on these prerequisites, various benefits of identifying key factors for success in the management of natural resources emerge since a significant percentage of SES around the world are managed by local communities:

1. A comprehensive quantitative model for understanding success in SES would be available for the first time.
2. New CPR projects (forest, fisheries, irrigation systems) could be designed or optimized according to the newly defined important success factors and their interactions.
3. The conditions of those people who suffer from problems caused by poorly organized institutions could be improved.
4. The failure of many projects could be explained by a practice-oriented clarification of success factors and, in the future, even be avoided.
5. A certain potential for reducing environmental pollution could be tapped worldwide. By using such a model and optimizing sustainable management,

the destruction of resources could be counteracted, and intergenerational justice could be increased.

6. The model could, at low cost, be implemented in any number of projects at the same time. A complete analysis of a SES could be standardized and be shortened considerably, provided that an independent data collection exists.

1.2.2 Gaps and obstacles

The question of factors relevant for SES success or failure is a central question in research on CPR. Accordingly, it has received much attention (Ostrom 1990; Ostrom, Gardner, and Walker 1994; Agrawal 2002; Hess 2008), but is still fraught with massive, unresolved problems. The following section discusses nine central problems. In addition to these nine main problems, there are other research gaps. They can only be mentioned shortly. These include, for example, mapping the causal pathways of the success factor's network. Hardly any study develops a model of the many causal factors, or even names more than one. Thus, only non-comparable individual studies with many weak correlations remain as basis for modeling. Another gap is a precise determination of the respective factor's importance for different classes of SES, the inclusion of a temporal component and the consideration of outside settings, for example, at a regional or national level. Yet another problematic factor is highly variable group-dynamic processes (Olson 1968; Wit and Kerr 2002).

We now come back to the nine central problems. These problems concern data (1–3), methodology (4–6), and theory (7–9):

1. *Data problem 1:* The first problem concerns the limitation of most empirical studies to the investigation of one or few cases. This problem often is due to the approach via case studies, which, as discussed previously, allow a detailed knowledge of a system, but do so at the expense of comparability. This is connected to the lack of a broad data base of comparable case studies.

2. *Data problem 2:* Since the case study approach is dominant (*data problem 1*), often only a few key variables are collected. This is usually justified through the thorough knowledge of the system and theoretical considerations. However, it is known that influencing factors may change their importance and even direction depending on other factors (here: unknown, because not collected). This is especially problematic when trying to generalize and transfer research results to other systems. Therefore, there is a large number of empirical studies that are limited to single systems with few (usually 2–4)

variables. Since, at the same time, there is a striking lack of large N-studies, comparisons are hardly possible (Poteete and Ostrom 2008; Poteete, Janssen, and Ostrom 2010).

3. *Data problem 3:* Existing case studies rarely agree on which factors are generally important or even whether they are positive or negative. Thus, for example, some studies find that a high dependency of users on the resource is positive since users assign it a correspondingly high value and maintain it well (Ostrom 2009). However, other studies contend that a high dependency, especially if survival depends on it, can lead to overuse or destruction of the resource (Agrawal 2007). Such differences are sometimes due to different methods or study design, but more often caused by success factors being operationalized differently (*methodological problem 2*). However, some factors are reported as statistically significant in some studies, whereas in other studies they do not show any effect. In still other studies they show a statistically significant effect toward the other direction (cf. e.g., Agrawal and Chhatre 2006; Cinner et al. 2012; Nagendra 2007).

4. *Methodological problem 1:* The fourth problem arises from the previously mentioned lack of comparability: the few attempts to make case studies comparable, such as meta-analyses, suffer from the loss of a lot of information since much study-specific information has to be transformed to a higher level of abstraction. In addition, the various different methods for collecting data have to be made compatible.

5. *Methodological problem 2:* Both concepts and variables are operationalized differently in almost every study (Frey 2017a). Therefore, the data basis in SES research is not uniform—no overall picture is able to emerge. There is hardly any comparability even if only one sector is concerned (e.g., forest management). Even key concepts such as social capital or environmental success are operationalized differently:

"This meta-study shows that measures of success discussed by the authors vary across all case studies. None of the selected articles discussed all measures of success simultaneously. This suggests that achieving all success's dimensions, that is, ecological, economic, and social at once, is difficult and complicated. Research settings, therefore, usually emphasize a certain dimension at a time to simplify testing models and ensure relevant outcomes to be observed." (Pagdee, Kim, and Daugherty 2006, p. 48)

Thus, it can be doubted whether success measures are defined the same way using these different ways. Especially dependent variables like ecological success are extremely difficult to measure in this respect. I will come back to this problem in detail in Section 4.2.6.

6. *Methodological problem 3:* A sixth problem is the lack of some kind of methodology to map non-linear, complex, and unknown interactions in a model. Conventional linear regressions, as used in most studies, are therefore only able to capture a fracture of actual interdependencies. To date, there is no quantitative model able to capture non-linearities and relationships between success factors of SES.

7. *Theoretical problem 1:* The seventh problem is that the interactions between potential success factors are largely unknown. Success factors are surely not simply additive. Individual studies provide some clues to a few complex and conditional interaction effects. This problem concerns, above all, interdependencies of possible factors with each other (Agrawal 2001). However, there is no suitable methodology for understanding the complexity of interactions in its entirety:

"These obstacles [to build a theory] exist in the shape of non-comparability of results from different studies, the problem of spurious correlation, and the difficulty of avoiding multiple and contingent causation in single case studies." (Agrawal 2001, p. 1665)

8. *Theoretical problem 2:* An eighth problem is the lack of an empirically grounded, comprehensive synthesis of success factors. Although a plethora of potential success factors which influence the sustainability of SES are mentioned in the literature, typically only a few factors are modeled (see *data problems 1* and *2*). Thus, despite some attempts (see Section 2.5), there is no comprehensive, systematic concept implementation of a synthesis of factors that is consistently based on empirical evidence. Without such a comprehensive base of independent variables (factors) any analysis can be but incomplete. Many researchers' decision to work with a small set of factors is, for pragmatic reasons, understandable. Nevertheless, partial sets of independent variables are not able to answer the key question— which factors are important, which ones are not? In order to clarify that question, a more comprehensive synthesis of success factors is needed, even if some factors will have a very low impact only (Pagdee, Kim, and Daugherty 2006).

9. *Theoretical problem 3:* All these difficulties result in another, last problem. There is no model of sustainable community resource management that contains all relevant success factors including their interactions. Although there is a generally accepted minimal set of important factors, a comprehensive model is yet unavailable. Hence, we still don't know whether success factors could be generalized across sectors. Is it possible to draw conclusions from, for example, irrigation projects' data to

solve forest management problems? Since most studies only examine projects of one category, cross-sector comparisons are practically non-existent. The number of potential success factors and the resulting complexity of dependencies have made comparative modeling impossible (Agrawal 2001):

> "Although much of this writing acknowledges the importance of a large number of different causal variables and processes, knowledge about the magnitude, relative contribution, and even direction of influence of different causal processes on resource management outcomes is still poor at best." (Agrawal and Chhatre 2006, p. 149)

Given all these problems, most notably no appropriate methodology, a synthesis of success factors that integrates three of the more comprehensive syntheses of factors (Ostrom 1990; Baland and Platteau 1996; Wade 1994) concludes pessimistically:

> "Using three of the more comprehensive such studies, and with an extensive review of writings on the commons, this paper demonstrates that the enterprise of generating lists of conditions under which commons are governed sustainably is a flawed and impossibly costly research task." (Agrawal 2001, p. 1649)

However, we do think that such success models are still possible and yield huge benefits. This book focuses on the solution of the nine main problems as a precondition for modeling. The following Table 1.1 again sums up the main obstacles as discussed previously.

In fact, a quantitative analysis of success factors is problematic to such an extent that some authors have completely questioned the success of such a project:

> At the same time, a large-N study of commons institutions that incorporated more than 30 independent variables and their interactions would require impossibly large samples and entail astronomically large costs. (Agrawal 2001, p. 1662)

Consequently, Agrawal proposed to reduce the variables to the most important ones. Which factors are to be described as "most important," however, depends on the subjective judgement of the individual researcher—with the result already being assumed. In other words, without an objective quantitative analysis, the different systematic data selection errors of the researchers (biases; examples are sample bias, conservatism bias, confirmation bias, among others, cf. Frey 2007) condemn the selection to be arbitrary.

Table 1.1 Summary of problems for establishing a quantitative model for ecological success in social-ecological systems

Number in text	Problem field	Problem
1	Data	Restriction of case studies to a few cases
2	Data	Restriction of case studies to a few variables
3	Data	Direction of factors is not robust and studies partially contradict each other
4	Methods	Methodologies in case studies are hardly comparable (despite meta–analyses)
5	Methods	Different operationalization of concepts, especially success
6	Methods	Lack of methodology to map non–linear, complex interactions between success factors in models
7	Theory	Interactions between potential success factors are unclear
8	Theory	There is no comprehensive conceptualization (synthesis of factors)
9	Theory	There is no quantitative, universally valid model; moreover, question of generalizability is still open

Even if such an analysis was not considered impossible, the difficulties mentioned led researchers to avoid undertaking the task. They regarded a comparative analysis of several large datasets with regard to the most important thirty to forty influencing factors as practically impossible. Thus, it has not yet been carried out.

Of course, there are meta-analyses (Pagdee, Kim, and Daugherty 2006; Gutiérrez, Hilborn, and Defeo 2011; Brooks, Waylen, and Borgerhoff Mulder 2012) which attempt to prove statistical correlations between selected factors of individual analyses. This, however, is satisfactory only to a limited extent since the real interdependencies between the factors remain unknown. Thus, any respective relevance of individual factors is hard to estimate (cf. for example Agrawal and Chhatre 2006): If, for example, a third factor that did in fact determine the outcome were omitted in the investigation of the correlation of two factors, then only a section of the real dependency of the three factors would be obtained (see *data problem 7 in Table 1.1*). This would inevitably create a wrong image of the present situation.

Here is how we plan to solve these problems in order to gain insights on success factors for the sustainable management of natural resources: we use a special data basis with a new methodology and an improvement of theory. Problems one to four are addressed by using three unique databases of SES case studies on irrigation, fisheries, and forest management. All in all, 794 case studies represent a broad data base for a robust and valid model. Comparability is given because case studies were collected for each database with the same methodology. Since for each data set between 550 and 800 variables are available, an extremely differentiated picture is possible. Details of the data can be found in Chapter 3.

The fifth problem (see Table 1.1), different operationalization of concepts, does not exist in this case since all databases were created by the same research group (the Ostrom Workshop in Political Theory and Policy Analysis). Thus, many variables overlap and are indeed based on the same theoretical background and empirical studies. This is particularly true since the theoretical analytical frameworks such as the institutional analysis and development framework (IAD), the design principles and the SES framework were developed on the basis of these data (Ostrom 1990; Poteete, Janssen, and Ostrom 2010). The disadvantage here is that the conceptualization stems from a single school of thought, which could mean a certain one-sidedness, for example, underestimating technical aspects.

Problems six and seven are solved by using machine learning algorithms, especially artificial neural networks (see Section 4.1.3). Neural networks are able to represent non-linear complex relationships. They are also able to find unknown structures in the data. This is particularly important because interactions between potential success factors are not known.

The eighth problem is addressed by a synthesis based on a literature review. We developed a comprehensive conceptualization out of all those potential success factors discussed in literature. The conceptualization now contains twenty-four success factors (see Section 5.1) which represent a systematic, theoretically sound synthesis of existing conceptualizations. With that, the last problem of generalizability can be tackled. Finally, the creation of three quantitative models and an overall model can answer the question of the general validity of success factors.

Hence, this analysis may indeed be groundbreaking: it implements a comprehensive, theoretically sound set of success factors that are operationalized consistently into a quantitative model based on a broad empirical basis of case studies. Today, to my knowledge, there is no quantitative model (but see for example Gutiérrez, Hilborn, and Defeo 2011 for a somewhat similar approach) capable of answering the above-mentioned questions concerning success and failure of SES in natural resources in this quantitative way. Having reviewed the main obstacles, it is now possible to spell out in detail the hypotheses in the following section.

1.2.3   Hypotheses

The following analyses are guided by three working hypotheses. They are derived from the main research question and the research gaps just discussed.

Hypothesis 1:
There are general models for ecological success in SES with at least satisfactory explanatory power, that is, there are general patterns for ecological success.

If combinations of success factors are specific for each case study, as sometimes postulated (Cleaver 2000), no general patterns should be found. Whether general patterns exist can be derived from the model quality, that is, the predictive accuracy. Of course, any threshold is arbitrary, but considering the complexity of factors, each model for ecological success in the field of SES, which explains more than 30% of the variance is a testimonial to the existence of such general patterns (Agrawal and Chhatre 2006). The results of such overall models can be found in Section 5.5.

Hypothesis 2:
Differences between ecologically successful and unsuccessful natural resource management can be largely explained by the selected success factors.

I assume that the synthesis of possible success factors can cover a large part of SES success's variance. This can be verified empirically by creating models—the twenty-four success factors differ between SES, so the ecological success should differ accordingly. As with the first hypothesis, model quality will decide upon rejection or acceptance of this hypothesis. A high accuracy of predicting the ecological success can only be achieved if all important factors have actually been included (see Section 2.5) and if they correctly describe the more general concepts (factors) (see Section 4.2). These results for the individual data sets are shown in Sections 5.2, 5.3, 5.4, and 5.6.

Hypothesis 3:
Neural networks are superior in their prediction quality to conventional statistical methods such as regressions, since they can take into account any presumed nonlinear and complex nature of interactions between success factors.

As described previously, interactions between success factors are largely unclear. Although there are a large number of studies that link two factors, such as social capital with compliance (Gibson, Williams, and Ostrom 2005), these relationships are often not statistically validated (Sandström and Widmark 2007), with the exception of a few isolated cases, but not systematically. Accordingly, no description of a network of interactions is available.

Thus, a method is needed that can independently create a model without theorizing, that is, without postulates about the type, direction, and strength of interactions between factors and success. Self-learning, supervised machine learning algorithms such as neural networks are able to do this. The results of this modeling and the comparison with other methods are presented in Sections 5.2, 5.3, 5.4, 5.5, and 5.6.

2

State of Research

This analysis combines several research areas. For the three key underlying areas, the current state of research must be discussed: the general structure of cooperation mechanisms in Biology, game theory research on social dilemmas in public goods, and the analysis of social-ecological systems (SES).

In the following, the state of research on these topics will be narrowed down to their significance for the research subject matter. This chapter (State of Research) first presents basic cooperation mechanisms and common-pool resource (CPR) problems of non-human beings as a basis for understanding (Section 2.1). Subsequently, the behavior of people in social dilemmas is discussed (Section 2.2). Social dilemmas are characterized by the fact that it is more advantageous for everyone involved to behave egoistically and to maximize their own interests. At the same time, the clash of such selfish strategies leads to a profit which is lower overall for everyone than it could be with behavior that does without maximizing one's own interests. The social optimum can only be achieved if all participants renounce short-term (rational) profit maximization.

Because of the enormous difficulties that influence factors on the willingness of people to cooperate in social dilemmas such as CPR management, the results of research from public goods games (PGG) in the laboratory are used. In a controlled environment, they provide valuable clues as to which factors and contexts in general lead to more investment in the CPR and which do not.

Then, the special characteristics of CPR systems are described and the underlying social dilemma about different theories such as game theory, cooperation theory, and rational choice theory is characterized in more detail (see Section 2.3). Against this theoretical background, the analytical framework for SES is developed (see Section 2.4). Special attention is paid to the institutional perspective of Elinor Ostrom (1990, 2009). The terminology used is defined. Finally, Section 2.5 looks at previous studies on possible success factors. Thus, all essential aspects of the research topic with regard to the state of the art are included. I will start with the state of research on biological cooperation.

Sustainable Governance of Natural Resources. Ulrich Frey, Oxford University Press (2020). © Oxford University Press.
DOI: 10.1093/oso/9780197502211.001.0001.

2.1 What are fundamental biological mechanisms of cooperation?

Before dealing with the characteristics of human social dilemma situations, biological conditions and mechanisms for the emergence and maintenance of cooperation must be discussed in more detail. For this purpose, I first analyze the emergence of cooperation by its evolutionary functional logic before describing the tragedy of the commons (see Hardin 1968) in various organisms (Rankin, Bargum, and Kokko 2007).

Cooperation is defined in Biology as: "a behavior which provides a benefit to another individual (recipient), and which is selected for because of its beneficial effect on the recipient." It holds: "cost and benefit are defined on the basis of the lifetime direct fitness consequences of a behavior" (West, Griffin, and Gardner 2007, p. 416). A prerequisite for cooperation is that living beings are able to cooperate at all. Cooperation can already be observed in bacteria (Velicer, Kroos, and Lenski 2000). The basic strategy options in interactions are called *altruism* (benefits for others at the expense of oneself) and *egoism* (benefits for oneself at the expense of others). All variations between these two extremes are found. In addition, the distinction between proximate (causal causes of action, "How does it work?") and ultimate (functional causes, "Where does it come from?") is essential (Tinbergen 1963; Voland 2013). A distinction is made between the following explanatory approaches for cooperation (see Table 2.1 according to Voland 2013; cf. also Nowak 2006)

In practice, it is almost impossible to separate the respective proportions from these different mechanisms. Therefore, none of these mechanisms can be excluded a priori if one of the many varieties of cooperation is investigated. In some cases, such as bacteria, which are often genetically practically identical, it is obvious to assume nepotism for cooperative behavior.

If a direct benefit for both actors is achieved, it is called *mutualism* (Sachs and Simms 2006; Clutton-Brock 2002). Free riders' behavior is not to be expected here. Examples of mutualism are coordinated hunting, for example in African wild dogs (Creel and Creel 1995) or spotted hyenas (Drea and Carter 2009). Also, many coalitions are the result of mutual cooperation, such as food sharing among monkeys (de Waal 1989) or joint chimpanzee hunting (Boesch 2001). Free riding, also termed cheating or defecting, is benefiting from the advantages generated by others without contributing to the costs.

Cooperation, on the other hand, can result from nepotistic altruism as well, which can lead to costs for the actor. Altruism in relation to genetically related partners is one of the most important mechanisms that makes cooperation possible (Dawkins 1976; Lehmann and Keller 2006).

Table 2.1 Evolutionary mechanisms for explaining cooperation

Scenario	Strategic goal	Operational logic	Conditions	Examples
Mutualism	Investment in collaborative behavior with cooperation gains for all parties involved without altruistic advance costs	Direct amortization of behavior with "accidental" benefits for third parties as well	Non–zero sum games	swarm formation, team sports
Biological market	Exchange of goods or services for personal use	Trade	Supply/ demand situations	Mating behavior, cleaner fish/client cooperation, "food for sex"
Reciprocal altruism	Investment in the well–being of a partner in anticipation of their response at a later opportunity	Exchange current fitness for later fitness	Longevity, trust, philopatry	Mutual support in emergency situations
Indirect (or strong) reciprocity	Investing in a partner you know is also investing in a partner	Exchange current fitness for later fitness	Trust, sanction of norm violators	Mutual support in emergency situations
Handicap-altruism (costly signaling)	Investment in one's own market value as a social or sexual partner by assuming altruistic costs	"Handicap principle": investment in communicative reliability	Prestige hierarchies, publicity	Big–game hunting, production of public goods, generosity, public donation behavior
Nepotistic altruism	Investment in a relative	Exchange direct fitness for indirect fitness	Common descent	"helper–at–nest behavior," family solidarity

Note: This table is taken from Voland (2013), p. 71.

Nepotism can be found throughout the animal and plant kingdom in all taxa, such as bacteria (Rainey and Rainey 2003), social insects (Trivers 1985; Wenseleers et al. 2004), and humans (Voland 2013). The explanation for this kind of altruism results from the consideration of the overall fitness of an individual. It is made up of one's own direct fitness and the fitness of relatives (indirect fitness), whose reproduction success is fostered by altruism (Hamilton 1964). This is described by the inequality r * B > C. Altruistic behavior is expected when the genetic relatedness r multiplied by the benefit B of the receiver are greater than the costs C.

A second mechanism is direct reciprocity (Trivers 1971; Schino 2007). This is done according to the motto "tit for tat." This mutual cooperation, which can also take place among non-relatives, is characterized by efficiency gains—the benefit for the parties involved is altogether higher than mutual defection. However, the incentive to defection remains intact, since reciprocity is typical of referring benefits to the other party in advance. This asymmetry can be exploited. This incentive is described, for example, by the prisoner's dilemma (see Section 2.3). So far, there is little evidence of this mechanism in Biology which makes it a controversial issue. One contested example is the division of food among common vampire bats (Voland 2013, p. 70). A remarkable exception to this are humans, who show global and complex cooperation, which is for example reflected in trade and markets.

The evidence for non-human beings to engage in indirect reciprocity is rather sparse (but see, for example, Bshary and Grutter 2006).

Indirect reciprocity also involves interactions with third parties that observe behavior and results of interactions (Leimar and Hammerstein 2001; Nowak and Sigmund 1998). Therefore, reputation is a crucial mechanism, because participants often choose future cooperation partners on the basis of a good reputation, that is, a high level of cooperation and a low number of defections (Wedekind and Milinski 2000; Barclay 2008).

A fourth form of cooperation is handicap altruism. Here, hidden qualities—mostly in the context of sexual selection—are communicated with expensive and therefore honest signals, because they are not falsifiable (Zahavi and Zahavi 1997; an overview is provided by Johnstone 1995). Besides magnificent feathers or a similar physical quality, altruism may be used as a honest signal. One speaks of a handicap if the following five conditions are fulfilled (according to Uhl and Voland 2002):

1. the social partner is elected according to signal expenditure;
2. the signaling partner incurs costs;
3. the signal causes costs for the transmitter and impairs fitness;
4. the signal indicates interesting hidden qualities;
5. effort and success are located in different areas of life.

A fifth form of cooperation is biological markets (Noë, van Hooff, and Hammerstein 2001). Biological markets are defined by supply meeting demand and "goods" or services being exchanged. This can be shown, for example, for cleaner fish offering their cleaning services to their clients, other fishes that may accept such offers or decline them (Bshary and Grutter 2002). Typical for biological markets are the possibility of fraud, rapid shifts in supply and demand, and asymmetric cost-benefit functions for stakeholders. All these approaches can be subsumed under inclusive fitness theory (Hamilton 1964; West, El Mouden, and Gardner 2011; West, Griffin, and Gardner 2007). Important mechanisms are partner choice and partner control.

Scholars continue to debate which mechanisms in which form are responsible for the fact that people are willing to cooperate. It is also controversially discussed how stable cooperation networks in many situations can emerge (Hagen and Hammerstein 2006; Wilson 2013; Guala 2012; Hawkes and Bird 2002). There exist win-win situations (mutualism), long interactions with reciprocal character, preferred cooperation with relatives and altruism as a costly signal, and cooperation mediated by choice of cooperating partners.

Another fundamental question is how cooperation can develop. Although this question cannot be clarified here, an attempt is made in the following to understand it better. In a hypothetical population consisting exclusively of defectors, the mechanisms of nepotistic altruism and the handicap principle are above all the means by which cooperation can gain a foothold. If groups are small, related, and therefore cooperative, this selection advantage brings the ability and willingness to cooperate into the world—and can then be rededicated, improved, and expanded.

This also applies to altruism, which arises through the handicap principle: Individuals create positive conditions for cooperation, since they bear the initial costs. Once altruism is established via a handicap, an arms race is, according to evolutionary functional logic, to be expected. More and more individuals are investing in bigger and bigger commitments until this behavior is stopped by boundary conditions. For example, social goods such as a guard system can be created, as Zahavi proves in the case of Arabian Babblers (*Turdoides squamiceps*; Zahavi and Zahavi 1997). Every guardian acts altruistically—he could, for example, look for food during the guarding period—but gains prestige through this and thus secures reproduction advantages.

Due to their exploitability, direct and indirect reciprocity are not likely to establish cooperation in a population dominated by defectors. It is possible that indirect reciprocity can even be interpreted as a mechanism that was only possible when necessary preconditions such as signaling pathways and cognitive devices for reputation were preformed by selection on handicaps (West, El Mouden, and Gardner 2011).

In order to fully understand cooperation problems both the mechanisms just discussed as well as the boundary conditions that influence the emergence and development of cooperation are important.

I start with boundary conditions that play a role for cooperative behavior in non-human living beings. The amount of the resource is important. If it is sufficiently available, the costs of cooperation are lower and the number of cooperators is higher, for example, in bacteria (Brockhurst et al. 2008). In Biology, the tragedy of the commons is rather rare, because it is effectively limited by the law of decreasing yields. This can be illustrated by using the example of groups: From a group's point of view, newly added individuals are unattractive when the group optimum for the benefit to be achieved is exceeded by adding new members. However, non-associated individuals benefit from group membership even after the group optimum has been exceeded, since the sub-optimal group benefit is usually greater than their benefit as a loner (Voland 2013).

Similarly, the exploitation of resources is usually the same: the more individuals, the lower the yield. As a result, more and more individuals migrate to other resources, resulting in a positive feedback-loop that can lead to the recovery of the resource. Highly fluctuating, time-shifted correlating predator-prey populations clearly show this effect. Cyclical dynamic equilibrium shifts in the strategies of the individual could also contribute to such a stabilization (Semmann, Krambeck, and Milinski 2003). It is individually advantageous not to engage in interactions in a population at all, if these are dominated by defectors (loners); but if there are many non-interacting individuals and therefore fewer defectors, then cooperation is worthwhile (again), especially in small groups. If the percentage of cooperators increases in turn, then the group advantage decreases because the benefit must be distributed to more individuals. In addition, defection is favored more strongly as the group to be exploited grows. As a result, the subgroup of defectors increases in strength, that is, this strategy becomes again more frequent. With many defectors the circle closes, because it becomes advantageous again to be a loner (see Figure 2.1).

This results in a fluctuating equilibrium, which is also known as "Rock-Paper-Scissors" and has been mathematically analyzed (e.g., Doebeli and Hauert 2005).

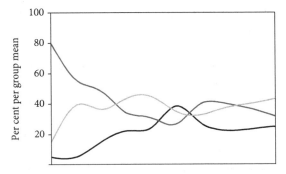

Figure 2.1 Interplay of strategies in a public goods game (black = loner, dark grey = cooperators, light grey = free rider; from Semmann et al. 2003)

Cooperation within groups must also be considered, since individuals often form groups (fish or flocks of birds, prides of lions, monkey hordes, and others). In humans, this is particularly pronounced. Beginning with families, villages, tribes, and ethnic groups, all the way to the level of nations—groups are cohesive units. Cooperating groups are thought to have an advantage over non-cooperating groups (Darwin 1874). Group selection as explanation must be refused (arguments for example in Dawkins 1976; Voland 2013).

Competition between groups, on the other hand, is frequent. Here, it is actually better for individuals to cooperate with the group under certain circumstances for selfish reasons. This mechanism is ecological displacement, not selection at group level. This mechanism promotes altruism, since cooperation makes efficiency gains possible. It is also possible to develop resources or social goods that would not be available otherwise (e.g., increased resistance to forces in bacteria through cooperation). A typical dilemma is created by territoriality. If a group defends certain resources, then their benefits are evenly distributed among the members, while the costs are not. They are based on individual participation in fights (for lionesses, see Heinsohn and Packer 1995; for guenons, see Mares, Young, and Clutton-Brock 2012). This may lead to selection pressures in the case of social primates (Willems, Hellriegel, and van Schaik 2013). Interaction partners have developed a number of mechanisms to prevent cooperation partners from not cooperating.

Perhaps the most important mechanism is punishment (Clutton-Brock and Parker 1995). Even in plants, there is monitoring of the symbionts' cooperation (Neuhauser and Fargione 2004) and punishment if not (Kiers et al. 2003). Also known in animals are many complex interaction patterns (Young et al. 2006; for fig trees and pollinating wasps, see Jandér and Herre 2010). Sanctions are also often observed on organic markets. Thus, for example, fish-clients punish cleaners for free-rider behavior with aggressive behavior (Bshary and Grutter 2002) or changing partners (Noë 2006). The question of the stability of such cooperation between species is the subject of much research (Leigh 2010; Sachs and Simms 2006; Raihani, Thornton, and Bshary 2012).

Some plants can use chemical sensors to measure whether a cooperation partner really cooperates. Animals also monitor whether they are deceived (e.g., ants, see Helanterä and Sundström 2007; honeybees or wasps, see Wenseleers et al. 2004; Ratnieks, Foster, and Wenseleers 2006; and birds, see Mulder and Langmore 1993).

Both free riders and highly cooperative individuals are actively selected according to their behavior in previous interactions. However, mostly humans show this selecting behavior, which is also called reputation or image scoring (Nowak and Sigmund 1998; Wedekind and Milinski 2000). In animals, there are only a few examples for such behavior (West, El Mouden, and Gardner 2011):

chimpanzees (Russell, Call, and Dunbar 2008) and fish (Bshary and Grutter 2006) have been observed to behave like that.

Reputation is above all a means of finding individuals who are not related and willing to cooperate. For relatives, this is also possible via markers, which originate from genetic kinship (for ants of the chemical species, see Helanterä and Sundström 2007) or are learned in childhood (for cooperative breeding birds, see Russell and Hatchwell 2001).

Spatial proximity within a group together with a spatial separation to non-cooperating individuals can promote cooperation. Examples are bacteria (Brockhurst et al. 2006; Kreft 2004; Donlan and Costerton 2002; Rainey and Rainey 2003).

With this brief overview of cooperation in non-human beings some basic parameters for cooperation in SES already come into view: size of the resource, group composition, size, spatial location, monitoring, punishment and also reputation play an important role (see Section 5.1.5). At the same time, the diversity of cooperation in the animal kingdom and in the underlying evolutionary mechanisms becomes evident.

A special problem class in cooperation situations is CPR. The following section describes in detail how CPR are treated in the animal and plant kingdom.

2.1.1 The tragedy of the commons exists for many different species

Cooperation problems arise not only in the exchange of services on biological markets or the coordination of activities such as joint hunting, but also in access to resources. Many natural resources have the character of a common resource (see Section 2.3), as they are easily over-exploited but difficult to delimit (Ostrom 1990). This is called the tragedy of the commons (cf. Hardin 1968).

Of course, it is possible to use only typical human factors as explanations for the tragedy of the commons (or its avoidance) in the case of CPR (see Section 2.5). However, it makes sense to search for general and possibly simpler mechanisms that can explain the phenomena just as well.

Therefore, it is advisable to start with simple organisms that are also confronted with CPR problems. There, defection can be understood as egotism that wants to receive benefits as quickly and directly as possible (related to itself). This usually implies a certain short-sightedness and the preference of actions that result in short-term, smaller profits compared to later, possibly larger profits.

Cooperation, on the other hand, is the result of an "enlightened" egoism, which also accepts situations in which the expected benefit arises either later (reciprocity) or directly or indirectly (e.g., via relatives). These cost-benefit calculations can be expressed biologically as fitness.

In fact, a tragedy of the commons occurs between many groups of organisms, even if the effects are often mitigated by various factors (Rankin, Bargum, and Kokko 2007). This problem is also observed within organisms, for example through the competition of genetic information. Other conflict situations relate to parent-child constellations or sexual competition (Rankin, Bargum, and Kokko 2007). Such a dilemma affects bacteria and plants that are in symbiosis with each other (cf. also Crespi 2001; Rainey and Rainey 2003). Other examples are social insects in which the interests of the hive collide with those of individual (Wenseleers et al. 2004).

Monitoring and punishment are two examples of mechanisms that can prevent biological collapse (overview in Rankin, Bargum, and Kokko 2007). These have been shown to occur in marine gastropods (Dall and Wedell 2005) and in many groups of social insects. These mechanisms also play a central role for human beings (see Section 2.5).

Reputation is another effective way of promoting cooperation in the animal kingdom. Reputation plays a central role in chimpanzees (Russell, Call, and Dunbar 2008; de Waal 1989), the Arabian Babblers (Zahavi and Zahavi 1997), and even in fish (Bshary and Grutter 2006).

Thus, chimpanzees observe cooperative behavior of others and adjust their behavior accordingly. Also, some fish can distinguish between cooperative and defective cleaner fish.

It is noticeable that none of the described CPR problems leads to a complete collapse, but only to a limited tragedy of the commons. However, this could also be due to the fact that collapsed resources can no longer be observed.

These observations on cooperation mechanisms in non-human living beings can be summarized as follows:

1. The benefit of cooperation is considerable, because despite the constant danger of exploitation by defectors, cooperation is observed throughout all taxa.
2. Behind all cooperative and egoistic strategies lie complex, sometimes long-term (non-conscious) cost-benefit calculations, which are checked by selection.
3. Higher cognitive abilities are not necessary for cooperation. Cooperation can be found in plants or bacteria. Cooperation partners must not be recognized and advantages of cooperation need not be noticed—selection alone is sufficient. Cooperation is therefore not exclusively to be understood as a social phenomenon.

4. The complete collapse of a resource is rare in Biology (Rankin, Bargum, and Kokko 2007; Kerr et al. 2006).

5. Especially among relatives, cooperation via inclusive fitness mechanisms is to be expected.

6. If animals live in groups, increased cooperation within the group is to be expected under certain conditions (e.g., small group size).

7. The group optimum for the exploitation of resources is often exceeded for selfish reasons of an individual (Voland 2013). What the individual considers to be a positive cost-benefit balance, however, may not be optimal from the group's point of view.

8. Both within groups and individuals, depending on the current composition and situation, there is often a change of strategy.

9. Reputation promotes cooperation in various taxa (e.g., birds and chimpanzees); it is thus a potentially significant mechanism for humans.

10. Surveillance and punishment effectively prevent the tragedy of the commons.

Based on these observations, the following section outlines how to increase cooperation rates.

2.1.2 Some possibilities for preventing resource overuse

In this section, a first summary attempts to collect possible approaches that can prevent resource overuse. First, the law of decreasing yields needs to be mentioned: a tragedy of the commons is often prevented with many species by a decreasing yield, combined with evasion, migration, or conversion (Rankin, Bargum, and Kokko 2007).

For humans, this means finding the point of optimal sustainable extraction and not exceeding it despite given technical possibilities. This is done, for example, by not allowing the appropriating group to grow indefinitely. In many communities there are mechanisms like restrictions on admission such as residency or long term affiliation to the local community (Ostrom 1992a).

If the yield of a resource decreases (e.g., a prey), more and more individuals substitute it and follow a cost-benefit curve (Krebs and Davies 1978). The resource recovers.

For example, fishermen often have protected spawning grounds and accepted bans to catch (heavy) females and young animals. Modern technologies such as large fishing vessels are problematic in this respect (Berkes 1986). They are able to sustainably damage populations with a single fishing operation. This can only be prevented if it is guaranteed that certain fishing areas are reserved exclusively for local fishermen and that the national fleets adhere to quotas.

On the other hand, there are many types of surveillance, punishment, and reputation. These measures are also very effective for humans and are described in Section 2.2.

In the animal kingdom, higher cooperation rates can be observed among relatives, which is also true for humans (Peters et al. 2004). Therefore, this is an approach for possible improvements. Strengthening group membership also increases the cooperation rate for both animals and humans.

Both proposals are, however, limited in scope (Voland 2003)—more advantageous, according to the previously mentioned study, is the use and creation of mutual interests for increasing cooperation in a globalized world.

It is also important to note that successful cooperation is highly context-dependent and ecologically specific. It is also subject to individual cost-benefit calculations. Both defection and the decision to cooperate depend on many factors, for example, the discount-curve, the relationship of the individuals, the group size, the current life situation of the individual, etc. This calculation is rejected or confirmed by means of natural selection (Voland 2013).

If the benefit is high enough, cooperation arises even without higher cognitive abilities, such as the symbiosis of nitrogen-fixing bacteria with plants (Kiers et al. 2003). This proves once again the importance of internal (and unconscious) cost-benefit calculations. The elucidation of these individual calculations among humans has high priority in order to be able to improve cooperation rates.

Since humans—at least theoretically—are in a position to overcome the relative short-sightedness of cost-benefit calculations of selection, the previously mentioned possibilities are within their reach. Scientific analyses and a certain overview of the situation make it possible to plan long-term and sustainably, for example via global climate and energy scenarios. The mentioned possibilities are, however, still too unspecific and unsystematic in this form. Therefore, they are tailored and specified in the following Sections 2.2 and 2.5 to social dilemma situations in the context of natural resources. The following is an overview of the basic influences in social dilemmas that influence people's cooperative behavior.

2.2 What drives cooperation in laboratory experiments?

The following sections (Sections 2.2–2.3) constrict the previous more general analyses to the two areas of research that can shed light on the research question (see Section 1.2).

This is, first of all, research on public goods dilemmas in the laboratory, since possible factors influencing the success of the social dilemma can be manipulated and controlled very well. However, the artificiality of the laboratory environment is problematic especially for the so-called lack of ecological validity, which has often been criticized (Wiessner 2009). For this reason, the present analysis is limited to an overview. This is, second, the explicit and more or less systematic attempt to identify success factors in SES (see Section 2.5).

One of the most influential attempts to understand cooperation is Game Theory (Nash 1951). It considers the decisions of all parties involved in the interplay. When examining their predictions—including Experimental Economics and evolutionary game theory—it has become evident that people often deviate from strategies considered optimal (Walker et al. 2000) and are not rational in the sense of classical economic theory (Henrich et al. 2001; Gintis 2000). Although some of these game-theoretical models provide useful predictions, they are at least incomplete, but possibly also erroneous in at least some of their assumptions. After decades of research, it has become clear that humans do not act according to game theory, as shown by the evidence given later in this section.

From an evolutionary biologically informed perspective, taken by the present paper, it is assumed that humans behave to a large extent optimally. The "deficits" discovered, such as individual, seemingly irrational strategies, are primarily due to ecologically inapplicable laboratory conditions and inadequate models that do not consider essential influencing factors. This means that game theory does not capture some important situation parameters relevant to human behavior.

This assumption of optimality is based on the fact that altruism, egoism, and varieties of cooperation based on them are fundamental evolutionary strategies (see Section 2.1). They occur in virtually all animal and plant species and constitute a complex of adaptations, on which there was considerable selection pressure. Non-optimal strategies have been and still are discarded depending on environmental conditions. Consequently, non-optimal behavior in ecologically valid situations can be optimal under laboratory conditions (Hagen and Hammerstein 2006).

In the following, empirical results from Experimental Economics relevant for CPR problems are presented. The results are presented primarily in terms of robustness, that is, whether they converge despite different experimental designs and methods. Thus, the problem of the lack of ecological validity is at least partially alleviated. It is then clarified whether these data can be brought into line with existing theories. The number of experiments and the variables studied, on which cooperation strategies and the contribution to public goods depend, is immense. The computer simulations also available for cooperation (e.g., Nowak and May 1992; Helbing and Yu 2009) are not discussed here in detail, since their advantages—many parameter changes with many repetitions—cannot outweigh their serious disadvantages, especially the lack of ecological validity. In my opinion, they are only of limited significance for real situations.

In contrast, field experiments (see Section 2.2.2) surpass laboratory experiments significantly in their ecological validity. This can be achieved by means of heterogeneous samples (not only students) and a context-specific formulation of the problem (Cardenas, Stranlund, and Willis 2000). Usually, a combination of a field experiment (behavior in a controlled environment, e.g., in the case of the exploitation of a CPR such as fish) and the actual behavior of the test subjects in their natural environment (as a fisherman; an example of such a design is Rustagi, Engel, and Kosfeld 2010) is used. Studies show that results from field and laboratory experiments are transferable in principle (Mitchell 2012). Nevertheless, there are not enough comparable data on individual parameter changes for field experiments, since they usually follow very individual experimental plans. The following overview is therefore ordered according to these influences (variables).

Among laboratory experiments, public goods games are the best choice as they represent the best compromise between mapping the CPR problem and availability in sufficient numbers and variations to determine the robustness of results. Moreover, the few experiments on CPR in which the yield function increases to optimum before it drops steeply (over-exploitation) confirm these results (Ostrom, Gardner, and Walker 1994). Results from ultimatum, dictator, and other games are also used for comparison.

In ultimatum games, a player must split an amount of money between himself and another player; if that player rejects the offered division, both go blank, if he accepts, both receive the proposed share.

In game theory, it is rational for player A to offer the smallest possible amount and for player B to accept it. However, this is not observed. Instead, often a substantial amount is shared. In dictator games, only the dividing player determines the distribution, without the possibility of vetoing player B. Here you would expect player A to keep everything to himself. Again, this is not observed (Henrich et al. 2010).

A typical public goods game presents the choice of the player to keep his or her endowment from the game supervisor or to contribute it to the public pool. The private pool is independent of the decisions of the other players and the return rate is 1:1. Public contribution depends on other players and a multiplier, with which the game supervisor rewards this choice. The only social dilemma is that it is rational for every player to do nothing (free riding). However, the social optimum and the highest efficiency are achieved if all players pay in full. In a typical PGG, players contribute about 40–60% of their bet in the first round of the public good. In the course of the game, this participation drops to a few percentage points (Fehr and Gächter 2000).

The next paragraphs describe the most important influencing factors on the willingness to pay in PGG into the public good, that is, the cooperation rate. These factors are arranged in ascending order according to influence and robustness of the findings—I start with factors whose influence is unknown. There are socio-biological explanations for the effectiveness of many of these factors, such as reputation or religiousness (e.g., Voland and Schiefenhövel 2009).

There are several reviews of this immense literature on PGG (for example Chaudhuri 2011; Ledyard 1995). Based on these studies, possible influencing factors were identified. The studies associated with the factors mentioned there have been analyzed by means of a literature search.

In addition, the sources mentioned there were followed up according to a snow ball system. Although the list of factors (Table 2.2) is certainly not complete, it covers all significant influences on cooperation behavior to my knowledge. For reasons of space, the factors are not discussed in detail, but only presented at a glance—with some important references included.

Although there are many other influencing factors considered in research, they are but rarely the subject of experimental manipulation. Therefore, at the current state of research, they can only be regarded as indications of their influence on human cooperative behavior. They are outlined subsequently.

Emotions

Emotions accompany many decisions or are their cause. *Free riders* evoke strong negative feelings (Dawes, McTavish, and Shaklee 1977; Fehr and Schmidt 1999). These feelings intensify the more free riders deviate from the group standard. There are also indications that cooperative actions are emotionally rewarded internally (for an fMRI [functional magnetic resonance imaging] study, see Rilling et al. 2002; for an explanation by a "warm-glow effect," see Palfrey and Prisbrey 1997).

Endowment

The willingness to cooperate also depends on how much "initial capital" is available. More capital leads to lower payments (Ostrom, Gardner, and Walker 1994). The efficiency decreases (from 99% to 73%). Both in the baseline (from 34% to 21%) and in the treatment with communication, which is an indicator for the readiness to cooperate.

End of game is known

Game theory predicts complete defection through backward induction when the end is known, since for "rational" players, in the last turn, defection is the only sensible option. Since this is known, the penultimate round becomes the last one. The same consideration applies to all other turns as well—until the first one. However, except for a few exceptions (international chess grandmasters, cf. Palacios-Huerta and Volij 2006) people do not behave like that in PGG.

Table 2.2 Overview of influencing factors in public goods games

Independent variable	Effect on cooperative behavior	References
Age	unknown, too little data	Gächter, Herrmann, and Thöni (2004); Egas and Riedl (2008); Fehr et al. (2003)
Heredity	unknown, too little data	Cesarini et al. (2008)
Educational level	unknown, too little data	Gächter, Herrmann, and Thöni (2004)
Intelligence quotient	unknown, too little data	Wilson, Near, and Miller (1996)
Sex	unclear	Koopmans and Rebers (2009); Nowell and Tinkler (1994); Cadsby and Maynes (1998b)
Children in comparison to adults	no significant influence	Harbaugh, Krause, and Liday (2003); Fehr, Bernhard, and Rockenbach (2008)
Socio-economic status	no significant influence	Gächter, Herrmann, and Thöni (2004)
Amount of real profits	no significant influence	Wiessner (2009); Cameron (1995)
Group size	no significant influence	Isaac, Walker, and Williams (1994); Milinski et al. (2006)
Instruction / Training	small to medium influence in both directions	Oosterbeek, Sloof, and van de Kuilen (2004); Isaac, McCue, and Plott (1985); Cadsby and Maynes (1998a)
Context sensitivity	strong influence in both directions	Barkow, Cosmides, and Tooby (1992); Kühberger (1998); Yamagishi et al. (2007)
Conditional strategies	very strong influence in both directions	Fischbacher, Gächter, and Fehr (2001); Kocher et al. (2008); Ones and Putterman (2007)
Learning and length of game	no significant influence / positive influence	Walker et al. (2000); Isaac, Walker, and Williams (1994); Gächter, Renner, and Sefton (2008)
Religiosity	positive influence	Sosis and Ruffle (2004); Atkinson and Bourrat (2011)
Homogeneous group	positive influence	List and Price (2009); Buchan, Croson, and Dawes (2002); Werthmann, Weingart, and Kirk (2008)

Table 2.2 *Continued*

Independent variable	Effect on cooperative behavior	References
Less anonymity	strong positive influence	Milinski et al. (2006); Franzen and Pointner (2012)
Market integration	strong positive influence	Henrich et al. (2001); Buchan et al. (2009)
Assortment	very strong positive influence	Gürerk, Irlenbusch, and Rockenbach (2006); Page, Putterman, and Unel (2005)
Fairness	very strong positive influence	Fehr and Schmidt (1999); Yamagishi et al. (2009)
Marginal per capita return MPCR)	very strong positive influence	Isaac, McCue, and Plott (1985); Brown–Kruse and Hummels (1993)
Country / Culture/ Level of globalization	very strong positive influence	Buchan et al. (2009); Ockenfels and Weimann (1999); Henrich et al. (2001)
Sanctions	very strong positive influence	Fehr and Gächter (2000); Nikiforakis and Normann (2008); Herrmann, Thöni, and Gächter (2008)
Reputation	very strong positive influence	Milinski et al. (2006); Barclay (2004)
Communication	very strong positive influence	Bochet and Putterman (2009); Ostrom, Gardner, and Walker (1994)

In more realistic, modified PGGs that are modeled after real CPR problems, however, the knowledge of impending destruction or a relatively large chance of destruction *always* leads to a very rapid, actual destruction of the resource (Ostrom, Gardner, and Walker 1994).

Cognitive stress

It has also been shown that participants break down more complex processes into simple rules of thumb. Subjects work with approximations for contributions and pay-offs. If they create new rules, they pay attention to simplicity. Heuristics are used to minimize the cognitive load of the working memory (Ostrom, Gardner, and Walker 1994). Typical examples of rules created in the laboratory are "Everyone gets the same share," "First come, first served," or "Everyone gets their share in a proportional manner."

Divisibility of the property

The divisibility of the public good, which can be attained through a joint group effort, also seems to be significant. The contributions for an indivisible good (82% of the stake) are almost doubled compared to a control group whose public good was divisible (43%) (Marwell and Ames 1981).

This is all the more important as all other manipulations in this study show almost no effects. One explanation is: The attempt to maximize divisible goods results in a 50% to 50% division of money for most players. In other words, players try to maximize both for themselves and for the group.

Non-divisible goods, on the other hand, are perceived as important and expensive because of their price, which leads to a higher investment (Alfano and Marwell 1980).

Deposit payment

Paying a deposit before playing a PGG makes cooperation the most sensible strategy (Gerber and Wichardt 2009). If not all participants deposit, the deposits will be returned before the game and a normal PGG will be played. If all participants have made a deposit, the PGG is played first and then deposits are paid out. Only those who contributed to the PGG get their money back.

This is similar to an early punishment; however, the sanction is known in advance, voluntary, done by oneself and therefore accepted. This has two advantages. First, it does not give rise to negative emotions or anti-social punishment, which is highly detrimental to efficiency (Nikiforakis and Normann 2008; Denant-Boemont, Masclet, and Noussair 2007). Secondly, no superordinate authority must exist or be accepted—a trustee is sufficient.

Laboratory situation

The laboratory situation is also a context factor (frame) and has an influence on observed cooperation rates (but see Koopmans and Rebers 2009, which cannot detect any difference between a normal and an online PGG). The subjects know that they are "tested" in some way. Thus, the results of anonymous experiments carried out on the computer might differ from experiments with an investigator. However, in attempts to guarantee strictest anonymity in PGG, there is no difference in the cooperation rate toward a control group (Laury, Walker, and Williams 1995). Still, this does not cancel out the basic test situation. These reflections on experimental situations mainly concern attempts to exclude reputational effects. This may be true for the experiment itself, but is usually led ad absurdum by the situation before it (welcome, instructions, pre-game testing, etc.). Participants know that their behavior is monitored and evaluated. Thus, reputational effects are not excluded in such situations with very high certainty. Through this superordinate situation (frame) unconscious behavior patterns and heuristics are called upon, which refer to the overall situation and cannot be switched on and off easily.

Two further points support this effect.

1. Experiments show that people are extremely sensitive to even the slightest evidence of surveillance (e.g., eyes instead of flower motifs on a money box, see Bateson, Nettle, and Roberts 2006).
2. A good reputation can only be achieved by acting in the same way in most, if not all, situations.

Therefore, individuals should also consider their reputation in situations in which they believe themselves to be unobserved, because the possible damage, if they should be observed unexpectedly, is immense (Frank 1988). The handicap principle explains why. It is expensive to show a certain behavior for years (God's faithful believer, badass businessman, etc.) and then have this reputation destroyed.

2.2.1 Bias and intercultural comparisons

Almost all of the sources mentioned so far suffer from distortion (bias). They examine Western students, mostly American psychology students. In the following, these results are therefore contrasted with intercultural studies. A difference is immediately obvious: There is a much wider range of behaviors—Western students show only a small part of it. They may even be completely atypical (Henrich, Heine, and Norenzayan 2010). An extreme example can illustrate this: The most frequently played offer in the dictator game with the Hadza, a people in Tanzania, is 0%, in the ultimatum game 10%, and in the "third party as a penalty game" again 0% (Marlowe 2009). Observations from the everyday life of Hadza confirm this picture. They are hardly capable of solving collective problems.

There are hardly any examples of long-term, collective work (e.g., fieldwork) with a fair division of labor, as frequent free rider repeatedly make them break down. There are few norms. In addition, violating these norms almost never leads to sanctions (Marlowe 2009).

The typical offer of Western subjects for ultimatum games is about 44% of the total. In fifteen small-scale companies from all over the world, the margin widens to 26% to 58% (Henrich et al. 2001).

On average, however, at 39%, they are close to the Western average again. PGG also show remarkable intercultural differences. Thus, contributions (with sanctions) range from 29% in Athens to 71% in Seoul and 90% in Boston. On average, they are at 65% (Herrmann, Thöni, and Gächter 2008). The previously mentioned PGG study (Buchan et al. 2009) also shows this wide range of the willingness to contribute to the common pool and ranges from 17% in Iran to 77% in the US.

Another study shows that Cambodian and Vietnamese rice farmers differ significantly under controlled and identical experimental conditions (Werthmann, Weingart, and Kirk 2008).

A comprehensive and consistent interpretation of this data is difficult because different studies test for different correlations with different designs. An explanatory complex of factors (market integration, level of globalization) can be described as *familiarity with trade*. An individual who is familiar with markets, exchange (reciprocity), and money should show a very high willingness to cooperate in daily life (Henrich et al. 2001; Buchan et al. 2009).

By contrast, a summary of the results of Ultimatum games played worldwide shows no major differences in terms of the level of contributions which are at about 40% (Oosterbeek, Sloof, and van de Kuilen 2004). However, the rejection differs regionally: Asians reject unfair offers more often, and even within the US, one finds differences, since Americans from the West Coast tend to reject unfair offers more often than those from the East Coast. Repetition has effects—the more often the Ultimatum game is played, the fairer the participants become, although the rejection rate remains the same. The size of the stake also plays a role (but see Cameron 1995). This effect is small, but it shows that the higher the stakes, the less money is contributed.

2.2.2 Field experiments

A principal question is the transferability of laboratory results to SES, that is, the lack of ecological validity. This is important because theoretical considerations from cooperation Biology and Economics are based on PGG in the laboratory context.

First of all, it is problematic that the majority of experiments are carried out with one particular group: American students in their first semesters. Whether this "sample" can be generalized to other cultures, age groups, and occupational groups has been disputed (Henrich, Heine, and Norenzayan 2010). At least in some aspects of cooperation, this group is rather atypical.

Another criticism concerns the *shortness* of interactions. A study can prove that both cooperation and punishment behavior change if the game is simply extended from ten to fifty rounds (Gächter, Renner, and Sefton 2008). This fact already points to the context dependency which is also reflected in many other aspects, such as the influence of the formulation and the framework (Yamagishi et al. 2007; Frey and Meier 2004).

Since there are results that show that subjects do not necessarily try to maximize their income only, for example, punishment in the last round, some researchers conclude that our cooperative abilities do not work properly in artificial environments. This is referred to as *mismatch theory* and mainly concerns individual interactions with anonymous, changing partners who cannot be assessed if they are cooperative or not since there is no information about their reputation (Zefferman 2014; Hagen and Hammerstein 2006). This suspicion is aggravated by findings that show that there are large differences between the behavior of test subjects in their natural environment and in artificial situations (Torres-Guevara and Schlüter 2016; Wiessner 2009; Levitt and List 2007; List 2006). Other experiments, on the other hand, have shown clear parallels between behavior in an artificial and natural environment (Rustagi, Engel, and Kosfeld 2010).

In the course of these debates, calls for more field experiments have become more pronounced (Rankin 2011). Although a general transferability from the laboratory to the field is given (Mitchell 2012), the differences between field studies, at least for problems with commons, are more pronounced than in the laboratory. While one study shows that control mechanisms to prevent over-exploitation of the resource function well (Reichhuber, Camacho, and Requate 2009), this does not occur in other environments (Cardenas, Stranlund, and Willis 2000). Similarly, the behavioral patterns range from strong egoism and virtually no cooperation (Stoop, Noussair, and van Soest 2009) to results in which individuals prefer the functioning of the ecosystem to personal profits (Cardenas 2000).

It is beyond question that people are in a position to draw up their own, adapted rules that promote sustainability (Cardenas 2000), even if it depends strongly on the nature and manner of decision-making mechanisms and participation (Cavalcanti, Schläpfer, and Schmid 2010).

As a conclusion, laboratory experiments provide important insights into human cooperative behavior in common pool situations; nevertheless, results must be interpreted with caution, since transferability and robustness are not always given, with various factors distorting the results. Field experiments try to build a bridge to the real world, but because of many compromises in experimental design, their results are often not as meaningful as laboratory experiments. Computer-based multiplayer environments that include social dilemmas in which the data of many players can be collected in an anonymous game situation may serve as a bridge between these two approaches (Frey 2017b). However, none of these environments can replace real case studies. The following sections (Sections 2.3–2.5) therefore focus on possible influencing factors in real case studies.

2.2.3 Behavioral experiments: Conclusion

Theoretical models: There are several attempts to summarize the diverse experimental results described in a theoretical model. Models based on simulations are problematic a priori (e.g., Boyd et al. 2003): The selection of influencing factors is based on unexplained and often unrealistic assumptions. In addition, only a few of them are selected. However, simulations may set empirical studies in motion. However, they cannot depict real complex processes reliably.

Another class of explanatory models is based on *experimental results*. Yet, none of these models ventures beyond two factors (e.g., fairness and economic environment, see Fehr and Schmidt 1999; niceness depending on intention, see Falk and Fischbacher 2006; punishment as implementation of strong reciprocity, see Carpenter et al. 2009). These models are therefore too simple to explain numerous and divergent results. In addition, the basis of these models is the rational, utility-maximizing human with clear preferences, as assumed by rational choice theory. This assumption is wrong, as numerous experiments prove.

Another point of criticism of existing theoretical models is their scope: The models relate exclusively to results obtained with Western subjects, often students, in the various paradigms such as Ultimatum, Dictator, and Public Goods games. However, as described previously, these results cannot be considered typical for humans in general (Henrich, Heine, and Norenzayan 2010). A generalization to other cultures is not possible. This is also supported by other intercultural studies (Herrmann, Thöni, and Gächter 2008).

That is why there is, to my knowledge, no model from experimental economy that describes the majority of empirical findings accurately enough and at the same time remains so general that it can be used for describing real CPR problems. For this reason, the discussion of theoretical models is limited to this minimal sketch.

Combination of factors

The abundance of experimental results can be used to profitably promote cooperation through appropriate parameter changes. This is described subsequently. First of all, it suggests itself to *combine* measures that promote cooperation. To my knowledge, there is no systematic investigation of this condition. Some studies combine influencing factors, but in different test conditions. Thus, there is no consistent picture. While the combination of punishment and reputation shows strong synergy effects (Rockenbach and Milinski 2006), the efficiency decreases with the combination of communication and punishment. Communication alone is closer to the social optimum (Ostrom, Walker, and Gardner 1992; Bochet, Page, and Putterman 2006). However, this is not always the case (Bochet and Putterman 2009). These results indicate that the influences cannot be added up linearly.

In addition, the individual measures between men and women seem to have different effects (Kurzban 2001), and the incidence of individual strategy types differs between countries (Kocher et al. 2008). At the same time, the results of some combinations of measures are intuitively understandable (e.g., the introduction of monitoring and sanctions, see Werthmann, Weingart, and Kirk 2008), while others are not (Carpenter 2004). Generally speaking, it can be said that there is a lack of research that systematically tests combinations for their effectiveness in promoting cooperation. Although no simple, linear correlations are to be expected, combinations of positive influencing factors should nevertheless reinforce each other in the majority of cases.

Summary
The analysis of economic experiments as a fundamental mechanism for influencing cooperation behavior serves as the basis for the factors of influence in SES that are discussed subsequently:

First, the described experimental results *indeed illustrate real behavior* with some significant limitations. This can be confirmed by data from real-life situations (Frank 1988). In the fictional emergency situation of a student collapsing in the subway, he/she is almost always helped (63 of 65 times); if he/she is visibly drunk, still about half of the people help. This is complemented by the fact that about 40% of lost wallets (in the US) are returned. Gambling behavior in trust games makes relatively consistent predictions of confidence in the real world, which can be verified using credit repayment rates (Karlan 2005). Also donations in two social funds of the University of Zurich prove that about 68% of students are willing to help anonymously without payment (which is similar to a dictator game, Frey and Meier 2004). In all these situations, *the context* plays a major role for the willingness to help.

Second, cooperative behavior and the efficiency actually achieved in attaining the social optimum is influenced by many factors. Simple models, which consist of just a few descriptive variables, cannot represent the complexity of human cooperative behavior.

Third, predictions of non-cooperative game theory do not apply: people are not egoistic maximizers, and they do not expect that of others (Bochet and Putterman 2009). In addition, egoists do not always best altruists. On the contrary:

> However, altruists seem to perform better economically: the experimental studies consistently show a positive correlation between altruistic behavior and socio-economic status. (Frank 1988, p. 195)

Fourth, the samples are too narrow and non-representative for many societies. Thus, any wider reaching conclusions about *the* cooperative behavior are not justified (Henrich, Heine, and Norenzayan 2010).

Fifth, in the case of slightly more complex CPR tasks (e.g., with a realistic convex yield function), test persons do not always understand where the social optimum lies at all. It is even more difficult to achieve the optimum with a group (Ostrom, Gardner, and Walker 1994). PGG are easier to understand with their linear yield function.

Sixth, typically, the result is neither 100% cooperation (social optimum) nor 100% defection (Nash equilibrium). Under realistic conditions (communication is possible, people know each other, problems can be contextualized and do not remain too abstract), the typical drop in contributions to the public good does not occur (e.g., Werthmann, Weingart, and Kirk 2008).

Seventh, punishment turns out to be an efficient tool in the long term (Gächter, Renner, and Sefton 2008; Frey and Rusch 2012). While many results contradict this hypothesis, this is very probably due to the fact that most PGG studies only run over ten rounds. Therefore, these results—especially the sharp drop in contributions in the last round and the resulting increase in sanctions—must be interpreted at least partly as laboratory artefacts. Sanctions decrease in the course of time, both in the laboratory and under real-world conditions, and a norm is established.

Eighth, context effects and game parameters play an important role. Test persons adapt their strategies partly flexibly to the circumstances and react to group composition, the amount of contributions of others, and many other contextual effects.

How can the these results be interpreted?

It is obvious to understand the positive effects of communication, homogeneous composition of the group, cooperation through assortment, and the high importance of fairness as facets of evolutionary adaptation. All this has been present in small groups for hundreds of thousands of years and probably represents the most typical and common case of cooperation among people. The negative influence of anonymity also fits in with this, which prevents information about all these variables.

Effectively, settings with the latter parameters turns players into strangers who are then initially met with mistrust. Short-term gains from defection are unlikely to be able to compete with the long-term efficiency of cooperation within small, known, and partially related groups (Wilson, Near, and Miller 1996).

Opportunistic behavior pays off when foreigners can be cheated in individual situations. The ability to discriminate precisely between altruists and egoists and consequently to select partners guarantees the highest profits. Even small differences should pay off here (similar to an arms race) and should be selected for, because even in small and well-known groups there are constantly different tasks to be solved with changing partners, each of which showing different commitment and cooperativeness according to the respective situation.

This is also supported by the fact that in the last round of PGG the defection rate is often increased. However, there are punishment and high contributions as well. From the point of view of game theory, this is irrational, incompatible with conventional models of Economics, and has so far been without convincing explanation. From an evolutionary point of view, however, this is understandable: it may be because people had to complete many different cooperative ventures with partially identical partners sequentially. Inconsistent behavior (cooperation, then defection, then cooperation again) prevents both development of trust and efficient cooperation in several projects. An evolutionary explanation for punishment in the last round is therefore that social interactions are not bound to individual situations, but to individuals (via reputation). In this case, it is consistent behavior to punish an interaction partner even towards the end: as a warning of possible defection in the next interaction.

Some other findings support the hypothesis that cooperation is a fundamental, biologically driven behavior. On the one hand, many reactions like punishment for revenge and free riders, building up reputation, and the positive internal evaluation of cooperation are controlled by emotions. On the other hand, learning effects seem to be extremely low (Andreoni 1988). If, after ten rounds, a new PGG with the same participants is announced, the restart effect sets in— participants show similar behavior to that of the first round (high contributions), even though they have already experienced that these investments do not pay off due to a decreasing willingness to pay. One possible interpretation could be: In real life, new situations constantly arise in which cooperation is necessary. It is therefore advisable to repeatedly demonstrate one's willingness to cooperate (which is reflected in 40–60% contributions in the first round of PGGs).

In my opinion, these findings allow only one conclusion. In spite of the constant danger of being exploited by free riders, it is better for individuals to cooperate by default in the long term in small, trusted in-groups, since many repeated cooperative interactions are win-win situations (mutual benefit) and can be explained by kin-selection or reciprocity. It is not necessary to use group selection models (see also Burnham and Johnson 2005). Both computer simulations and laboratory experiments point towards this direction. Important mechanisms are communication, reputation, and the possibility of punishment or changing partners. Good altruist and egoist recognition go hand in hand with this. These biological mechanisms, most of which also play a major role for successful cooperation in non-human species (Rankin, Bargum, and Kokko 2007), are the basis for a number of success factors (e.g., communication for factor concerning information, reputation for factor concerning social capital, punishment for factor concerning compliance.

2.3 Common-pool resources

2.3.1 Characterization of common-pool resource problems

Among the many varieties of natural resource management, CPR stand out in particular, as this form of ownership and management is very common worldwide:

- About 11% to 18% of the forests are commons (Chhatre and Agrawal 2008; White and Martin 2002).
- About 98% of all fishermen work in small local fisheries and catch more than half of the world's fish (Berkes et al. 2001, p. vii).
- Irrigation systems are responsible for about 90% of the world's drinking water consumption (70% of total fresh water consumption, see Siebert et al. 2010). Irrigated agriculture produces about 40% of the world's agricultural produce and is dominated by small farms (90%, see Cifdaloz et al. 2010).

However, the CPR problem structure is not limited to managing natural resources. On the contrary, it is a global and extremely important class of problems. It is found in animals and plants (Rankin, Bargum, and Kokko 2007) as well as in humans.

In addition to traditionally studied CPR problems such as irrigation systems, forest management, and fishing, there is also a wealth of new commons: Hess distinguishes between global, traditional, medical, and cultural commons, as well as commons in terms of infrastructure, neighborhood, knowledge, and markets (Hess 2008).

Another approach distinguishes between *existing* resources (e.g., forests), social goods that only *emerge* through cooperation (e.g., blood banks), and social goods, which arise as a result of *restraint in exploitation* or possible conflicts (e.g., global exemption from customs duty) (modified according to Rankin, Bargum, and Kokko 2007).

CPR problems have been given a lot of attention across disciplinary boundaries. Many approaches, including game theory (Nash 1951), Evolutionary Biology (Voland 2013), social science (Olson 1968), and Economics (Ostrom 1990; Fehr et al. 2003; Ostrom 2009) have contributed to solving this complex of problems. Because of their importance and the particularly acute social dilemma that is inherent, the present analysis focuses on this form of management of natural resources. An interdisciplinary approach based on biological cooperation research is chosen.

Ostrom offers a precise definition of CPR:

The term "common pool resource" refers to a natural or man-made resource system that is sufficiently large as to make it costly (but not impossible) to exclude potential beneficiaries from obtaining benefits from its use. (Ostrom 1990, p. 30)

The following definition also includes both the question of sustainability and the dilemma situation:

A commons is a resource shared by a group where the resource is vulnerable to enclosure, overuse and social dilemmas. Unlike a public good, it requires management and protection in order to sustain it. (Hess 2008, p. 37)

Three essential features characterize CPR problems: On the one hand, it is costly but not impossible to exclude potential beneficiaries of the resource. This is often due to the size and type of the resource (fishing grounds, pastures, or forests). On the other hand, the resource is limited, that is, each resource unit that a participant takes out is no longer available for other users (Ostrom 1990). Finally, although a group of collaborators is superior to a group of defectors at the group level, each collaborator is subject to exploitation by defectors at the individual level. With this clarification of terms, CPR can be distinguished from other related problems like public good problems. Such a demarcation is shown in Figure 2.2.

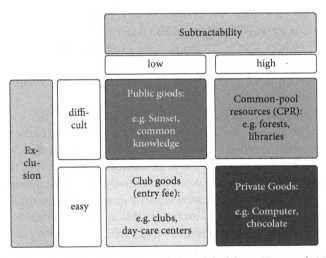

Figure 2.2 Characteristics of public goods (modified from Hess et al. 2003, p. 120)

Public goods are prime examples of social dilemmas. They are characterized by the fact that they are difficult to divide but easy to access (Hess and Ostrom 2003). Examples are sunsets or general knowledge. It is practically impossible, or very costly, to deny people access to a sunset.

At the same time, a sunset cannot be split. A public good can theoretically also be produced by a single person (e.g., construction of a lighthouse), while cooperation trivially requires at least two participants.

In contrast to this, CPR are easier to divide with the same difficulty of excludability. Their main difference to public goods is that they can be destroyed by overuse or pollution (subtractability). This is the case, for example, for clean air, parks, or the world's fish stocks. Sunsets cannot be overexploited. Goods that can be easily divided and whose individual access is also easy to regulate are generally privatized (e.g., consumer goods). If a good is difficult to divide and the access is easy to control, they are called club goods (e.g., pay-TV).

It should be noted that some goods which are usually classified as public goods can also be exhausted. It should also be noted that institutional measures can move a good within this matrix. A car can serve as an example. Usually it is a private good and belongs to an individual alone. Car sharing, on the other hand, allows a car to be shared among several paying members—the car becomes a club good. Unusual, but also possible is to provide one's car to the public free of charge—so that it becomes a publicly accessible resource (CPR) as it can only be used by one person at a time. If a state provided so many cars for free that there was no longer any rivalry, cars would become a public good.

In any analysis it is also important to understand the constituents of the common good. In a forest, for example, a CPR problem might occur when it comes to removing wood, but it could also occur with regard to the recreational function it provides. Figure 2.2 can provide a rough orientation only, since its boundaries often overlap and several problems intertwine.

The dilemma may also not arise at all, which is another possibility not captured by the above matrix.

In fact, for natural resources, this possibility is practically non-existent anymore. Theoretically, however, it is possible that the interaction of the technology used with a certain number of users remains below a critical threshold despite selfish and exploitative strategies. Thus, sustainable regeneration of the resource can take place. Since both the extraction technology and the number of users today exceed the natural regenerative capacity of virtually all natural resources, this possibility is almost always excluded. This applies a fortiori because rich natural resources typically lead to increased exploitation and rapid population growth.

After this classification, I will now deal with the social dilemma underlying CPR problems. The description of the basic strategic options with regard to cooperation is provided by game theory (non-cooperative game theory). Game theory is a mathematical description of the abstract possibilities for the clash of different individual strategies (Nash 1951).

The actual types of behavior and strategies of cooperation of human beings in such situations are analyzed in Experimental and Behavioral Economics (e.g., Fehr, Fischbacher, and Gächter 2002). Strategies here refer to the options of an individual regarding cooperation, that is, whether they cooperate, which means investing in the public good (altruism), whether they make this investment dependent on the behavior of others (conditional cooperation, Fischbacher, Gächter, and Fehr 2001), or whether it applies a free rider's strategy, i.e. nothing is contributed to the public good (egoism). It has been demonstrated that a considerable number of people cooperate in experiments, that is, invest in the common good (Gächter, Renner, and Sefton 2008). In Experimental Economics the dilemma of public goods consists in *creating* a public good by contributing to it by those individuals involved. This is usually done by depositing resources (usually money) provided by the experiment leader into a "public good" and simulating the efficiency of the cooperation of test subjects by doubling the contributions (Chaudhuri 2011). This extension of the prisoner's dilemma to more than two people and more than one round is called a *PGG*. Game-theoretically, a PGG is called an *iterated prisoner dilemma* (Axelrod and Hamilton 1981).

CPR games are much rarer in contrast to PGG (Anderies and Janssen 2013). Here, over-use is simulated and leads to the collapse of the resource if extraction is too high. In such a case, the resource does not recover.

There are many other experimental implementations of social dilemmas besides the PGG. The best known are trust, ultimatum, and dictator games (Güth, Schmittberger, and Schwarze 1982). The great advantages of these experimental implementations are the simple manipulations of individual parameters, the excellent theoretical understanding of mechanisms, and the clear results in response to manipulations of individual parameters. The CPR problem most closely resembles the so-called PGG or the CPR game. For the latter, however, hardly any data is available. I therefore concentrate on the numerous findings in PGG.

The possible strategies of the players (cooperation and defection) and their interaction are shown as wins and losses for each player.

$$T \text{ (temptation)} > R \text{ (reward)} > P \text{ (punishment)} > S \text{ (sucker's pay-off)}$$

Equation 1. Pay-offs in the prisoner's dilemma 1
And in addition:

$$2R > T + S$$

Equation 2. Pay-offs in the prisoner's dilemma 2

The highest profit can be achieved by defection, encountering, and exploiting a cooperative strategy. This results in the pay-offs T and S. If both players choose defection, this results in P for both. If both cooperate, this results in R for both. This cooperation gain R is greater than the pay-offs T and S together (winnings of the defector (T) encountering cooperation (S); see also Equation 2). Defection leads to a Nash equilibrium, as this strategy is optimal *regardless* of the opponent's strategy. This means that in CPR situations, the dilemma of the individual actors between egoism and altruism is particularly acute:

> Social dilemmas occur whenever individuals in interdependent situations face choices in which the maximization of short-term self-interest yields outcomes leaving all participants worse off than feasible alternatives. (Ostrom 1998, p. 1)

Most CPR problems are based on such or similar pay-off distributions. They can be modeled as repeated N-persons prisoner dilemmas (PGG) with the logic just described. However, repetition results in other strategic possibilities, since it is possible to react to strategies of the other player. Cooperation is therefore risky, but it can be worthwhile in the long run if it is reciprocated. However, it is more advantageous for every participant in CPR situations to maximize their profit by defection (appropriation without restraint), since each unit taken from the jointly managed resource is no longer available for others.

Defection in the real world is usually synonymous with damage to the resource. Since CPR situations are characterized by finite resources, that is, susceptibility to exploitation, this strategy choice by all or most of the participants leads more or less quickly to the collapse or the forced abandonment of the resource. An efficient, long-term use for the best of all is therefore only possible through stable cooperation. This situation is a *social dilemma*. This means that with cooperation, the short-term profit is lower than with defection. At the same time, however, there is a constant risk that profit-maximizing individuals exploit cooperation offers exclusively for their own benefit. A purely economic approach to natural resources usually leads to over-exploitation and has often been observed (Ostrom 1990). Biological CPR situations, by contrast, tend to show dynamic equilibria that prevent over-exploitation through negative feedback processes (Rankin, Bargum, and Kokko 2007).

In contrast to idealized laboratory conditions, there are, of course, a number of differences to real CPR cases:

1. Players usually cannot earn a reputation, and communication is typically not possible.

2. Contributions and pay-offs do not have to be comparable; for example, one co-operation partner may contribute labor while another may contribute money.

3. Contributions and pay-offs do not have to be symmetrical; it is unlikely that exactly the same contribution or pay-off will occur. Individuals expect different values and often assess the situation differently.

4. Contributions and pay-offs do not have to be measured in absolute terms; often they are valued by the parties involved, their pay-offs in relation to each other, and not in absolute amounts (Hsee et al. 2009). Hsee et al. (2009) specify that it depends on the type of good, more precisely, its implicit comparability (see, however, Diener et al. 1993, for example, which do not find support for this hypothesis of relative importance).

5. Not all strategies are rational in the sense of game theory. Strategies can also be based on heuristics (e.g., imitation), instincts, habits, emotions, or simply misjudgments or ignorance of the situation. A wide range of work has been carried out in the field of Biology (e.g., Frank 1988; Axelrod and Hamilton 1981; Krebs and Davies 1978). Further factors are described in detail in Section 2.1.

In PGG, significant differences in the level of cooperation, that is, the efficiency and success achieved, are particularly evident. Success changes if external environmental parameters are varied, for example, by punishment (Fehr et al. 2003), reputation (Rockenbach and Milinski 2006), or communication (Ostrom, Gardner, and Walker 1994). This indicates possible success factors. Biological research emphasizes that cooperative decisions through selection are tested strategies that are often long-term and can react flexibly to changed cost-benefit calculations (Voland 2013). Social sciences (e.g., Olson 1968) and Institutional Economics (Ostrom, Gardner, and Walker 1994) see the design of institutions as crucial. Above all, these parameters allow individuals to be persuaded to cooperate, despite incentives to the contrary. In addition, endogenous parameters must be taken into account, such as the choice of cooperating partners and the resulting group dynamics (Page, Putterman, and Unel 2005).

The previous sections (Sections 2.1–2.3.1) defined and characterized the basic structure of CPR problems. The following section (Section 2.3.2) reinforces this characterization of the dilemma situation.

2.3.2 The common-pool problem structure—exemplified

The following—predominantly biological—examples are selected to cover the widest thematic spectrum possible (cf. Hess 2008 and the Digital Library of the Commons, http://dlc.dlib.indiana.edu/, for a comprehensive literature overview).

Example 1: Tree height growth

Trees (participants) compete with each other for light (resource). Every tree invests energy in its growth (cost) in order to get as much light as possible for itself (defection respective self-interest). If all trees were to limit their growth to 1 m, all trees would reach the same height. This would correspond to the cooperation case. This is obviously not the case and is connected—due to the evolutionary selection mechanism— with a lack of communication and high incentives for defection (the latter are factors for failure). This situation is a zero-sum game with decreasing productivity (Rankin, Bargum, and Kokko 2007), since profit and loss arise from the difference in height, not by the absolute height. Absolute height, however, costs resources (thicker trunks) that are lacking in other areas. One observes decreasing productivity—in addition to that, collapse is to be expected as soon as a certain point is reached.

Example 2: Irrigation

Farmers (participants) in dry areas need water to irrigate their fields. To do this, they invest in an irrigation system (resource). Excessive water withdrawal (defection) often means lack of water due to drought. Fields are left for other farmers as too dry. In addition, the canal system must be maintained (costs)—here, non-participation corresponds already to defection. Only if fair distribution is guaranteed (requiring rules), is cooperation achieved. Poor controllability, inappropriate, poorly chosen rules, and lack of sanctions for defectors make stable cooperation difficult. A breakdown consists in the abandonment of the irrigation system. This happens if it has not been successfully maintained, or if those participants who have received no or little water intentionally destroy it.

Example 3: Traffic jams (internet, roads)

Traffic participants (participants) want to get from A to B in the shortest possible time (resource).

The following description applies analogously for data packages in the internet, that is, grid problems in general. If all participants select the fastest route and not the assigned or load balancing route (defection), this leads to overload and waiting times (costs) or even traffic jams (collapse). Cooperation is achieved when the participants are distributed approximately optimal, that is, with minimal delays on all available routes. This is often not done because a precise prediction of the behavior of other participants is too complex a task. In addition, there are enormous difficulties of information transmission to all participants, even if algorithms for such situations exist.

Example 4: Antibiotic resistance

People suffering from infections and animals sometimes even without infections (participants) are often treated with antibiotics (resource). Too widespread use or premature discontinuation (defection or self-interest) of treatment (in animals and humans) leads to antibiotic resistance (Cars, Hedin, and Heddini 2011). Resistant bacterial strains can no longer be destroyed with this antibiotic (collapse). Completed treatments and appropriate dosage in animals correspond to cooperation. This is aggravated by a lack of knowledge, disdain for the consequences, and lack of control.

Example 5: Climate change

Humans (participants) cause climate change with the combustion of fossil fuels (Edenhofer et al. 2011). If at least the majority of people continue to show climate-damaging behavior despite better knowledge (defection or self-interest) and does not switch to (more expensive) replacement solutions such as solar panels, better isolation, etc. in the short term (costs), there will be a major climate change with negative consequences until collapse. International cooperation and individual ecologically oriented behavior corresponds to cooperation. Obstacles are short-term profit orientation, national economic interests, and disdain for the consequences.

The examples outlined here seem to have nothing in common except for the abstract CPR problem structure. Thus, this small selection—CPR problems in organisms, natural resources, infrastructure, medicine, global politics, and information technology—illustrates the omnipresence and spread of this general problem of cooperation. Indeed, CPR problems can be found in all areas and at all levels. They occur everywhere where living beings encounter limited resources which, when harvested or used too heavily, show a drastic decrease in yield function until collapse. Since the range of actors involved is also very broad—ranging from bacteria to trees to people—it seems appropriate to search for general success factors while considering the respective structure of the situation and the boundary conditions which, too, influence the interacting strategies of the participants. Thus, these factors in their combination ultimately cause success or failure of the entire system. These complex interactions are analyzed in Section 2.4.

2.4 A primer on social-ecological systems

2.4.1 Three very different ways to manage common-pool resources

Three approaches to the management of natural resources can be distinguished—government-based management, privatization, or management by local user groups. A look at existing institutions in SES quickly clears up two prejudices. On the one hand, local user groups are able to use collective resources sustainably—collapse, that is, the tragedy of the commons, is not inevitable. Successful examples include groundwater supply in California and irrigation systems in Sri Lanka and Spain, fishing cooperatives in Turkey, and meadows as commons in Switzerland and Japan (Ostrom 1990). The last two examples exist since the Middle Ages and can be described as very stable.

Because of their obvious success, local institutions must be considered next to the two common solutions for public goods mentioned previously—nationalization and privatization (Ostrom 1990). It depends on many context-specific factors which of the three solutions is optimal. In addition, mixed forms are conceivable and do exist. They are discussed in Section 2.4.3.2.

According to today's state of research, an a priori preference for only one solution, as it is common in Economics in particular, is not supported by results from biological cooperation research and New Institutional Economics (Ménard and Shirley 2008). Some general advantages and disadvantages of these three solutions are obvious:

A state solution usually has excellent (centralized) control options. It is an external and therefore neutral "arbitrator." Therefore, resources are often allocated fairly. Boundary costs remain low as the resource is retained in its entirety (Hardin 1993). In addition, the government has the expertise needed to exploit the resource close to the sustainable optimum. However, this is an ideal image—none of these advantages must necessarily exist.

Typical disadvantages are corruption, as well as high costs for controls, sanctions, and information gathering (Agrawal and Gibson 1999). High costs cause capacity problems, especially with larger resource systems like large irrigation schemes; conflicts of interest between different objectives of the state are possible as well. Often it is impossible to guarantee effective monitoring. Moreover, the state or the inspectors employed have often little motivation and adequate local knowledge to intervene in the best possible way. This leads, for example, to errors in the severity of the punishment (Ostrom 1990).

When *privatizing* CPR, the use of resources should be optimal out of self-interest—but only ideally (Ostrom et al. 2002). In reality, it depends on the owner's knowledge, skills, and resources. In addition, privatization of public goods can lead to expropriation of existing users. A strategy of short-term profit maximization, such as the privatization of parts of the Bavarian state forest, can also lead to over-exploitation. Another disadvantage is the small scale. It entails high costs for large purchases such as implements and multiple expenses for protection (e.g., fencing). Small-participation often leads to non-optimal distribution of resources (Baland and Platteau 1996). This results in under-utilization and unproductivity, as well as other problems, for example because over-utilization cannot be compensated by rotation or similar measures. This problem is also referred to as *the problem of the anticommons* (Schurr 2006). The most serious restriction is probably that some public goods such as fish stocks can be either hardly privatized or not at all.

Through their knowledge, *collectively acting local institutions* in their use of resources can also operate close to the optimum of sustainable extraction, as it is postulated for privatized goods. The costs of monitoring, sanctions, and obtaining information are distributed among the members and can be minimal (Baland and Platteau 1996). In an ideal case, everyone is motivated to find optimal rules since every member is interested in using the resource. Like the state, local institutions benefit from eliminating the costs of demarcation (e.g., fencing) and the possibility of a balanced resource use (e.g., migrating shoals of fish, rotating grazing, etc.).

Disadvantages are the enormous difficulties of developing efficient, fair, and sustainable rule systems. That is what many cooperatives fail to do. The result of such a failure was described by Hardin (1968) as the *tragedy of the commons*. As already mentioned previously, Hardin considers only situations where exclusion is not possible (open access), that is, his arguments are not about CPR problems (e.g., traditional commons) in the strict sense. Moreover, only a few—mostly negative—relevant factors exist. So it is no surprise that this special constellation of factors fails (Ostrom 2007). Therefore, the question of a more comprehensive view of the causes of success in SES is crucial.

2.4.2 Institutional analysis of social-ecological systems by Elinor Ostrom

Around Elinor Ostrom a network of theoretical and empirical studies that analyze SES has been developed. As described in the previous section (Section 4.1), the theoretical background is based on New Institutional Economics (Williamson 1975; Hagedorn 2008; Ménard and Shirley 2008), in particular transaction cost theory (Coase 1937; Coase 1960) and the theory of public goods.

On the one hand, a decisive progress has been achieved through theoretical advances in the form of frameworks, and on the other hand through operationalization of many theoretical concepts and their empirical examination of case studies. The two analytical frameworks developed—the Institutional Analysis and Development (IAD) framework and the SES framework—will be presented in Section 2.4.2.2. Operationalization takes place through the development of three consecutive case study databases, in the course of which several hundred variables have been developed over a period of thirty years. This allows concepts such as social capital, the emergence of institutions but also biophysical attributes of a resource system to be recorded precisely (see Sections 3.1, 3.2, and

3.3). The decisive factor is the consideration of very diverse aspects, either bio-physical, group, rule or system-specific attributes.

Thus, this research sets itself apart from the development cooperation of past decades, which primarily focused on technical solutions (cf. Lam 1998; Ostrom 1992a) as well as purely institutional research exclusively focused on regulatory systems (Gruber 2008).

Ostrom's preparatory work (1990) has branched out and started a network of researchers (Cox, Arnold, and Villamayor Tomas 2010). This is particularly true in the area of data sets (see Section 3) made available to other researchers in form of additional case studies (data set "CPR": for example Nilsson 2001; Frey and Rusch 2013). Much attention has also been paid to issues of optimal institutions in irrigation systems (data set "NIIS": Lam 1998; Tang 1992; Shivakoti and Ostrom 2002; Joshi 2000). Finally, there are the many different activities on sustainable forest management that have also resulted from the development of a worldwide database by Ostrom (data set "IFRI": van Laerhoven 2010; Gibson, Williams, and Ostrom 2005; Chhatre and Agrawal 2008; Wollenberg et al. 2007; Andersson and Agrawal 2011). The terminology used by this network is described in more detail in the following section (Section 2.4.2.1).

2.4.2.1 Some conceptual clarifications

SES are the core concept. Contrary to the prevailing viewpoint, for example in development cooperation, which often considers mostly technical improvements as relevant, it also draws attention to the social components of a system and not only to the physical conditions (Anderies, Janssen, and Ostrom 2004).

Many examples demonstrate that it is crucial to involve local user groups in collective decision-making processes (Berkes 2007; Pagdee, Kim, and Daugherty 2006). Despite the typical view that the construction of a new dam as the core of an irrigation system is a task for specialized engineers, many projects can be decisively improved by the local knowledge of the users (Ostrom 1992a; Tang 1992). In such a view, collective action is about decisions in, for example, infrastructure. However, a related approach sees institutions as one type of infrastructure (Anderies, Janssen, and Schlager 2016).

It is important above all that the integrated viewpoint of this approach includes social, ecological, and economic aspects in its analysis on an equal footing. This is evident, for example, in the topmost level of the SES framework model: resource system and resource units are two categories that take up important ecological system attributes while actors and governance systems do so for social and economic attributes (Ostrom 2009). Due to the anchoring in New Institutional Economics there is a slight emphasis on institutional attributes. Recent work in this area, however, is trying to strengthen the role of ecological attributes (Vogt et al. 2015; Glaser et al. 2012).

Historically, the distinction between *open access* and *CPR management* is significant for the development of theories of SES. Garret Hardin's influential analysis (1968) rightly comes to a pessimistic conclusion about sustainable management of open access resources. Elinor Ostrom points out, however, that this is only valid for open access cases. However, the majority of natural resources are managed as CPR (Ostrom 2007). For this purpose, she developed a number of factors that are important for self-organization, that is, describing the transition from open access to CPR management (Ostrom 2007). It can be shown that although the exclusion of unauthorized persons is difficult and expensive, it is nevertheless possible. In addition, extraction restrictions and similar rules must be developed to prevent over-exploitation. This results from the rivalry of natural resources.

As discussed in Section 2.3, there are a number of problems in CPR management if the community's interests do not coincide with decisions that are most rational for the individual. These social dilemmas are often referred to as *collective action problems*:

> The very concept of a collective-action problem implies a disjuncture between individual preferences over outcomes, strategically rational individual behavior, and collectively desirable outcomes. (Poteete, Janssen, and Ostrom 2010, p. 67)

Without appropriate regulations, there are incentives for free riders not to participate in investments in the common good. Three situations within CPR management are particularly vulnerable: provision, distribution, and appropriation.

For the *provision* of a public asset, such as irrigation systems, work is required. For example, the sewer system must be kept in order and maintenance work must be carried out. If not enough individuals of the community cooperate, the common good cannot be made available. In the case of the *distribution* of water, which usually follows certain rules—such as rotation by time, sequence, or proportion according to field area—it is possible to gain personal advantage at the expense of the community by not complying with the rules. Finally, appropriation is usually also precisely regulated: fishermen may only catch a certain quantity of fish (quota systems), which can, however, be easily circumvented if monitoring is negligent.

Rules can thus defuse this dilemma and lead individuals to contribute to the collective good.

However, they must be monitored and enforced. Ostrom distinguishes between three central types of rules: operational choice, collective choice, and constitutional choice rules.

Operational rules directly affect day-to-day decisions made by the participants in any setting. These can change relatively rapidly—from day to day. (Ostrom 2005, p. 58)

They are not necessarily formally adopted and may change relatively quickly if necessary (details are shown in Baur and Binder 2013 for Swiss common pastures). However, they are—especially when other rules are missing—the de facto standard according to which action is taken.

Often the so-called collective-choice rules exist as well:

Collective-choice rules affect operational activities and results through their effects in determining who is eligible to be a participant and the specific rules to be used in changing operational rules. These change at a much slower pace [than operational rules]. (Ostrom 2005, p. 58)

Actual behavior is formalized by them:

The collective-choice rules define how individual "votes" at an operational level are aggregated toward a decision to add a new rule to the social-ecological system. (Poteete, Janssen, and Ostrom 2010, pp. 190–191)

Collective-choice rules are typically designed, established, amended, or deleted at annual general meetings or similar committee meetings of the respective stakeholder groups.

Finally, there are so-called constitutional-choice rules:

Constitutional-choice rules first affect collective-choice activities by determining who is eligible to be a participant and the rules to be used in crafting the set of collective-choice rules that, in turn, affect the set of operational rules. Constitutional-choice rules change at the slowest pace. (Ostrom 2005, p. 58)

They thus fulfil a kind of legislative function and are of decisive importance for success (Madrigal, Alpízar, and Schlüter 2011). They also describe who is allowed to participate in the development of rules. Equipped with these terms, the analytical frameworks used can now be characterized in greater detail.

2.4.2.2 A short historical account of different frameworks analyzing social-ecological systems

A comprehensive view of success factors requires a theoretical analysis framework, since success factor syntheses, such as those presented in Section 2.5, only represent a scaffold of system attributes description. In addition, many other conditions in which SES are embedded must be observed. Frameworks try to capture this. An analytical framework contains the most general variables and tries "to identify the *universal* elements that any relevant theory would need to include" (Ostrom 2005, p. 28).

Models, on the other hand, make precise assumptions about a limited set of parameters and variables. Due to the complexity, different levels of analysis are necessary. Ostrom uses as analogy the different resolution levels of maps for different analysis purposes. Thus, for example, in *Governing the Commons*, the costs of rule changes and situation variables that influence decisions of the institutional election are compiled as analytical frameworks (Ostrom 1990). In this analytical framework, which is presented subsequently, the decision-making process (Figure 2.3) begins with the recipients' assessment of the current situation with regard to any rule changes. Can the CPR be split? Can the resource be protected from unauthorized use? How can the CPR be divided up? This depends on the physical attributes of the resource, the technical possibilities for extraction, the characteristics of the community, and the existing rules.

The answer to this question depends on the kind of information available to the recipients or whether and how easily (cheaply) information can be obtained. Modified by community norms, the time factor of implementation, and the expected future benefits, costs and benefits of changes are calculated. This is done in an institutional framework (action arena) created by the actors. In this action arena, players take positions such as chairman of an arbitral tribunal.

The costs can be broken down into costs ex ante and ex post. The changes themselves include costs (ex ante). They depend, among other things, on how previous decisions have been made, on how able the leaders, are and how autonomous the community is to change rules (for a complete list see Ostrom, Gardner, and Walker 1994). The costs ex post consist of monitoring and enforcement of agreed rules.

The type of rules and the legitimacy or the possibility of excluding others participants play an important role. Expected costs are compared with expected benefits. Benefits are calculated, among other things, on the basis of the CPR size, the number of actors, the type and number of conflicts, and the current state of the CPR.

All these variables are used for decision-making. This can be done by a few decision-makers or by everyone concerned. That is, the variables represent information that may or may not be available. Often, information is scarce. On the basis of this information, each affected party calculates costs and benefits for themselves and supports or rejects rule changes. Rule changes relate, for example, to the appropriation from the CPR, the organization to implement rules, or to deal with external actors such as the state.

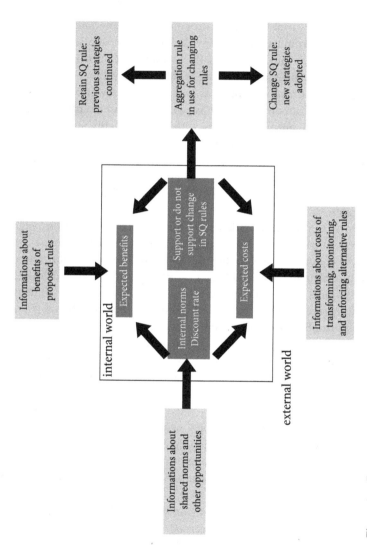

Figure 2.3 Decision-making process to establish new institutional rules (modified from Ostrom 2005, p. 193)

The frequently used IAD framework model (Figure 2.4), on the other hand, zooms out in detail from this view and describes in an abstract and very general way the entities involved in institutional decisions and their relationship among themselves (Ostrom, Gardner, and Walker 1994; Ostrom 2005; Oakerson 1992).

The interdisciplinary nature of the framework model is clearly visible. Thus, economists tend to deal with rules, while environmental scientists tend to deal with the biophysical attributes of the system, and social scientists tend to deal with group attributes (cf. Anderies and Janssen 2013), although many studies exist that use the entire model.

The biophysical conditions in the IAD framework model are usually referred to as boundary conditions, that is, restrictions for social interactions, which changes in the SES framework that is based on the IAD. There they are considered attributes, not boundary conditions. In the case of community attributes, the heterogeneity of individuals is usually taken into account. This concerns, for example, economic, ethnic, and educational differences.

The rule system (rules-in-use) distinguishes between rules that are actually valid (de facto) and rules that are officially valid (de jure) since it is possible that these two rulesets may not be identical or even contradict each other. Norms and moral values are also considered here (Ostrom 2005).

These conditions finally influence the actual acting individuals (participants) in a situation (action situation). An action situation is usually associated with several other action situations (interactions). An action arena, on the other hand, takes into account the fact that different individuals in different positions create different results using the same structured action situation (Anderies and Janssen 2013). An action arena is defined as follows: "An action arena occurs whenever individuals interact, exchange goods, or solve problems" (Anderies and Janssen 2013, p. 17).

For example, farmers interact in an irrigation system both during rule-making and in harvesting situations and work to maintain the system—all of these action situations are interrelated (interactions). Many analyses are now primarily interested in how to evaluate the outcome of such institutions and their rule systems (evaluative criteria). A methodological problem is to investigate the determinants of success and how they are measured (see Section 4.2.6).

The IAD framework is also characterized by the solid theoretical foundation behind the supposedly simple visualization (see Figure 2.4).

Thus, participants can be in different positions. A farmer can be the chairman of some water user association with extensive rights for rule changes, while he is a farmer among many when it comes to water allocation. It is not possible to go into any further details regarding the positions, rules, and interactions between the components of the IAD framework model here (see Ostrom 2005; Anderies and Janssen 2013).

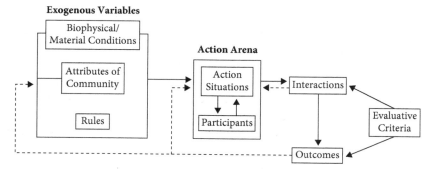

Figure 2.4 IAD framework for the description of SES (from Ostrom 2005, p. 15)

For the present analysis, the IAD framework model provides the basic structure that allows to capture and compare various institutional interactions. This flexibility and the abstract selection of layers also characterize the further development—the SES framework—that is used in this work.

The SES framework is the most recent and perhaps the most comprehensive framework in terms of levels (Ostrom 2007, 2009). It is also by far the most influential (Partelow 2018; Thiel, Adamseged, and Baake 2015).

Figure 2.5 illustrates how the interactions of the action situation from the IAD framework model have remained in the central area, towards which everything is focused.

New, however, is the division into four subsystems—resource system (RS), resource units (RU), rule systems (GS), and users (U)—each of which has a number of important attributes with which they influence the results through the interactions (I). The framework is structured in tiers: the top level (first tier) comprises the subsystems mentioned previously, the environment (related ecosystems), and the social, economic. and political framework conditions. The second tier represents the attributes for each subsystem, such as RS1 (size) for the resource system.

However, the tiers are not consistent. Thus, the concepts for social, economic, and political settings are at a national or global level, while—at least partly—some biophysical attributes are at a very specific, individual level (e.g., RU6). The interactions span a very wide range of scales as well. Furthermore, some attributes are very specific (for example RS3, RU5, U9), while others subsume a whole field of subdivisions and additional specifications (e.g., GS5–7, RS6). However, all are on the same level.

Furthermore, Elinor Ostrom emphasizes the nesting of individual components. Thus, several rule systems can affect one or more resource systems at the same time, and the default case is that several groups of actors exist side by

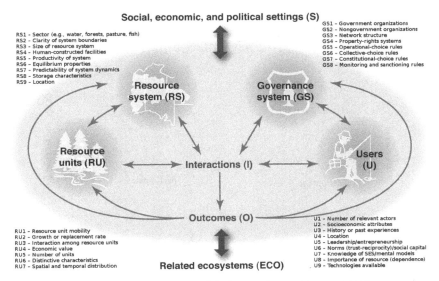

Figure 2.5 SES framework model (modified according to Ostrom 2009)

side in an SES. This is also illustrated in Figure 2.6, as each subsystem is drawn in several instances.

One of the objectives of the SES framework model is to keep an eye on the complexity and non-linearity of SES. At the same time, it is clear that unambiguous operationalization of thirty-six system attributes (plus eight interactions) is already a difficult undertaking (Hinkel et al. 2015). A further, explicit goal is the systematization and support of data collection. Its abstract level makes it possible to use and adapt it to very different SES, from dryland areas (Addison and Greiner 2016) to fisheries (Basurto, Gelcich, and Ostrom 2013) to water management (Hileman, Hicks, and Jones 2015). Furthermore, there are some attempts to adapt it to related scientific fields like the ecosystem service approach (Rova and Pranovi 2017) or sustainability research (Partelow 2015). The present analysis uses the highest level of the SES framework model to systematize success factors. Another use is made of system attributes since they are incorporated into the success factor synthesis (see also Section 2.5, factor synthesis 10).

The SES framework model itself has evolved since 2007. Two innovations are already visible in Figure 2.6. Thus, the renaming of users as actors takes into account the fact that many participating groups do not have to be users. The first name was historically due to its origins in natural resource management. The subsystem Resource Units (RU) has been renamed as well.

It is now called Resource Services and Units (RSU). Here, too, an extension has taken place because resource systems produce not only units but also

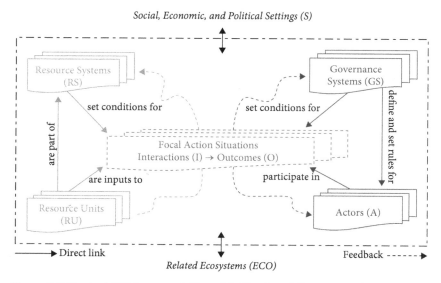

Figure 2.6 Revised SES framework (from McGinnis and Ostrom 2014)

services (ecosystem services). These can be immaterial, such as the aesthetics of a forest.

In addition to the SES framework used, there are many other frameworks (e.g., the Management and Transition Framework, see Pahl-Wostl et al. 2010). An overview article (Binder et al. 2013) classifies twelve frameworks for SES, including the SES framework. Classification is based on three criteria: firstly, whether the framework model is static or dynamical; secondly, whether it adopts an ecocentric or anthropocentric perspective; thirdly, whether it allows to abstract from certain SES and analyzes different types or not.

The SES framework model is classified here as static, anthropocentric, and general. It does not allow an explicit process analysis over time, thinking mainly of the user's influence on the resource system and is applicable to different SES. As discussed previously, the SES framework is used in this analysis because it is tailored to fit the case studies and is easily operationalizable. In addition, the empirical data used here have been developed in conjunction with this framework (Ostrom 1990, 1992a; Wollenberg et al. 2007).

2.4.3 Other related research approaches

After the approach of Elinor Ostrom was presented (Section 2.4.2), the superordinate theoretical approaches are briefly characterized in order to ensure classification. The analysis of SES is less influenced by an environmental-economic approach, which aims to set appropriate market incentives for dealing with natural resources. Resource-economic and ecological-economic approaches see themselves as systems sciences with focus on sustainability, but their emphasis is clearly on economic aspects. This is not the case here. SES research is primarily based on research on social and ecological issues (Ostrom 2009; Cox 2010; Berkes, Colding, and Folke 2003). This field of research sees solutions to sustainability problems above all in the theoretical and practical, that is, transdisciplinary linking of ecological, social, and economic aspects (Ostrom 2008).

Using a perspective that considers natural resource management to be an interaction of biological and social factors (Ostrom 1990), institutional aspects play a prominent role. How do actors work together? What rules can they use to prevent free riders from taking advantage of the system? How do they manage to design institutions in an ecologically sustainable, economically efficient, and socially just manner?

Such a perspective is also taken up by New Institutional Economics. As several authors point out, New Institutional Economics consists of a broad and diverse group of neighboring themes (Klein 1998; Williamson 2000). They share

> mostly criticisms of orthodox economics: (1) a focus on collective rather than individual action; (2) a preference for an "evolutionary" rather than mechanistic approach to the economy; and (3) an emphasis on empirical observation over deductive reasoning. (Klein 1998, pp. 456–457)

In this critique of neoclassicism, for example, transaction costs are underlined (Coase 1960; Hagedorn 2008), the role of social attributes such as social capital is emphasized (Pretty 2003), and it is pointed out that classical assumptions such as complete information are not realistic (Williamson 1975) and evolutionary mechanisms of action are relevant.

Humans are perceived as beings who are limited in their cognitive capacity and who do not always make rational decisions. In doing so, older social psychology concepts are taken up, such as bounded rationality (Newell and Simon 1972). Thus, human decision making is considered as not always using optimal heuristics. Heuristics are short-cuts in problem-solving that are not optimal but suffice to reach a goal. In addition, the latter have to operate under capacity and time constraints. The New Institutional Economics therefore rejects the concept of *homo economicus* and accepts distortions of (economic) decision-making, such as the anchor effect (Chapman and Johnson 2002), the framing effect (Kühberger 1998), or the sunk cost bias as part of our cognitive apparatus. In principle, decisions are taken under uncertainty (Gilovich, Griffin, and Kahneman 2002; Williamson 1975).

As the name suggests, institutions are regarded as central to the understanding of many processes, within SES. One focus is on institutional change (Schlüter 2007). Although there are many definitions of institutions, both broad and narrow, the following two are particularly suitable for characterizing them:

> institutions are enduring regularities of human actions in situations structured by rules, norms, and shared strategies, as well as the physical world. (Crawford and Ostrom 1995, p. 582)

A broader definition is as follows:

> Institutions are the written and unwritten rules, norms and constraints that humans devise to reduce uncertainty and control their environment. These include (i) written rules and agreements that govern contractual relations and corporate governance, (ii) constitutions, laws and rules that govern politics, government, finance, and society more broadly, and (iii) unwritten codes of conduct, norms of behavior, and beliefs. (Ménard and Shirley 2008, p. 1)

Institutions become effective through various structures of governance. Governance can be understood as actors (Stoker 1998) who make institutions effective through organizational structures and procedures (Hagedorn 2008). Governance has an impact, inter alia, on legislation, economic incentives, and other governance mechanisms (Jentoft, van Son, and Bjørkan 2007). In New Institutional Economics, the following governance structures are usually distinguished:

> Organizational arrangements are the different modes of governance that agents implement to support production and exchange. These include (i) markets, firms, and the various combinations of forms that economic actors develop to facilitate transactions and (ii) contractual agreements that provide a framework for organizing activities, as well as (iii) the behavioral traits that underlie the arrangements chosen. (Ménard and Shirley 2008, p. 1)

Other forms can be added to these, such as information, measurement methods, conflict resolution, and incentives for innovation. An analysis from the perspective of New Institutional Economics is typically complex, since it combines different theories (for an example see Schlüter 2006).

In addition to the preceding theoretical classification, the following sections (2.4.3.1–2.4.3.2) provide an insight into related, adjacent fields of research, on the one hand on *resilience* and on the other hand on *co-management* of SES, because important cross-connections exist here, such as the ecosystem character and biological aspects. By presenting commonalities and differences, the choice of analysis instruments is motivated once again; by broadening the perspective, instrumental limits become more visible. There is also a study providing an overview about theoretical linkages between theories within these related research fields (Cox et al. 2016).

2.4.3.1 Resilience of social-ecological systems

A further adjoining area of research is concerned with resilience and robustness of SES (Berkes, Colding, and Folke 2003). Based on the realization that ecological systems have different, stable equilibria (Holling 1973), the ability of a system to remain in a stable system state despite external shocks and the nature of the transitions between these equilibria is analyzed in more detail. Resilience is defined as follows:

> Resilience is the capacity of a system to absorb disturbance and reorganize while undergoing change so as to still retain essentially the same function, structure, identity, and feedback. (Walker et al. 2004, p. 2)

If it is differentiated between various types of research on resilience (Folke 2006), the link to SES research becomes clear. It is true that in many resilience studies, the main focus is on the resilience of (larger) ecosystems and the description of various biological equilibria and shock resistance, but there is also the explicit application to SES as I have defined them. Here, resilience is rather equated with the ability to adapt to major changes and the ability to learn (Pahl-Wostl 2009), which in turn has an impact on sustainable management: "Adaptability is the capacity of actors in a system to influence resilience. In a SES, this amounts to the capacity of humans to manage resilience" (Folke 2006, p. 2).

What can natural resource management learn from this perspective? On the one hand, it is the perception of the natural resource itself that must be considered as a dynamic flow equilibrium, but not as a static system that is automatically in equilibrium (Folke et al. 2002). On the other hand, this line of research makes explicit reference to the enormous difficulties involved in optimizing management, since complexity and dynamics of SES are extremely high (Folke 2006). Another problem is identified by the fact that different subsystems are on different scales, from very small to very large (Folke et al. 2002).

Thus, this approach leads to some recommendations for the management of natural resources (Folke et al. 2002): First, it reinforces the emphasis on the entanglement of social, ecological, and economic aspects in SES. Second, it emphasizes the flexibility of institutions as a central building block for necessary adjustments (Herrfahrdt-Pähle and Pahl-Wostl 2012). Third, an important role in adaptation processes is assigned to the respective local opportunities for participation. Fourth, embedding in social networks helps to support flexible institutions and develop appropriate rules (Folke 2006). Fifth, the biophysical attributes of SES and its dynamics, play a central role. This puts into perspective the importance of factors related to governance, as analyzed by Elinor Ostrom, especially in her direct proximity (for example Gruber 2008).

The article "A Framework to Analyze the Robustness of Social-Ecological Systems from an Institutional Perspective" (Anderies, Janssen, and Ostrom 2004) represents an important intersection of research on resilience and SES. Here, too, the focus is on the question "what makes SES robust?" and answers it by trying to create similar success factors for robust systems in analogy to the design principles. However, only the original design principles are discussed with regard to robustness. The analytical framework developed there is more influential, identifying resource systems, resource users, public infrastructure, and public infrastructure providers as the four most important subsystems. In addition, the most important interactions between these subsystems are identified together with the problems typically arising, such as free riding. However, this analytical framework is too rough for the purposes of this analysis. The lack of concepts and levels makes operationalization impossible. The authors themselves point out this problem.

Conclusion: In this field of research, potential success factors such as social capital, relationships with other actors, and opportunities for participation, flexibility of institutions, constant willingness to learn, and information about the ecological status of the resource are considered to be important. I will return to some of these success factors in Section 5.1. Above all, the willingness to constantly call up new information on the ecological status of the resource in order to adapt flexibly to changing conditions is the focus of another, adjacent research area as well—that of adaptive co-management, which is described in the following section (Section 4.3.2).

2.4.3.2 (Adaptive) co-management of social-ecological systems

Research on the resilience of SES brings the biophysical properties of systems to the fore, and so does the so-called (adaptive) collaborative management approach for the cooperation of local communities with superordinate organizations such as the state (Pahl-Wostl 2009; Plummer and Armitage 2007). One of many possible definitions of co-management is therefore

> the sharing of responsibilities, rights and duties between the primary stakeholders, in particular, local communities and the nation state; a decentralized approach to decision making that involves the local users in the decision-making process as equals with the nation-state. (Carlsson and Berkes 2005, p. 66 as defined by the World Bank)

If adaptive management (Costanza 1998; Berkes, Colding, and Folke 2003) is merged with co-management approaches (Plummer and Armitage 2007), the term adaptive co-management can be defined as follows:

> a process by which institutional arrangements and ecological knowledge are tested and revised in a dynamic, ongoing, self-organized process of trial-and-error. (Folke et al. 2002, p. 8)

This integration was at least in part anticipated in the 1990s by Kai Lee (Huitema et al. 2009). Co-management has evolved from the insight that neither non-regulated natural resources (open access) nor top-down management, usually by governmental authorities or organizations, have led to sustainable management as a rule (Gutiérrez, Hilborn, and Defeo 2011). On the contrary, the situation has often deteriorated as functioning, adapted local solutions have been replaced with standard approaches (Agrawal and Gibson 1999; Cinner et al. 2012). Instead of such one-sided approaches, co-management relies on genuine cooperation between state and local actors. Thus, co-management sees itself as a counter-draft to centralized, bureaucratic, top-down decision-making processes (Armitage et al. 2009).

A distinction is made between seven levels depending on which group of actors has more decision-making power (according to Sandström and Widmark 2007, again modified according to Arnstein 1969):

Level 1—Information: the state informs local users about decisions already taken

Level 2—Dialogue: the state enters into dialogue with users; this usually happens late in the process without users having decision-making powers

Level 3—Communication: Communication takes place between the two parties; local issues are included in management plans

Level 4—Advisory Bodies: Users get first decision-making power; usually, decisions are not yet binding

Level 5—Cooperation: Users can make limited petitions within the decision-making process

Level 6—Joint management bodies: Users have rights in bodies that go beyond an advisory role

Level 7—Partnership: equal rights of parties with institutionalized, joint decision-making processes

This hierarchy already suggests that in practice it is extremely difficult to determine exactly what kind of form cooperation has actually taken. For this reason, some authors also suggest that co-management should be analyzed as a process rather than just look at formal structures (Carlsson and Berkes 2005). In addition, it should be noted that the idealized case of two partners, a homogeneous user group and a coherent organization of a state, virtually never occurs. Instead, various actors are embedded in a network in which they seek solutions through joint deliberation processes (Pahl-Wostl et al. 2007; Carlsson and Berkes 2005). Other dimensions exist as well (Berkes 2009). There have been explicit attempts to link SES research to co-management (Stevenson and Tissot 2014).

The concepts of resilience and adaptable co-management are especially linked in research on adaptive co-management (Plummer and Armitage 2007; for an overview see Huitema et al. 2009). In addition to a secure livelihood for the affected individuals, sustainability is the overriding goal (Plummer and Armitage 2007). While co-management focuses on the actual rights of local actors in their interaction with the state (Pomeroy, Katon, and Harkes 1998), adaptive co-management is more about the adaptability of institutions (Plummer and Armitage 2007). In co-management, it is above all a matter of determining the rules in an open communicative and collaborative process, whereby the rights of the local community are of particular importance (Jentoft, McCay, and Wilson 1998).

Particular strengths of both approaches are seen in problems that affect several organizational levels (*vertical scaling*, for example state actors, regional control authorities, and local users) and are distributed geographically (*horizontal scaling*; see Armitage et al. 2009; Berkes 2007).

Co-management can also analyze decisions under uncertainty and in conflicts (Armitage et al. 2009), as the involvement of most actors helps to avoid unilateral "solutions," since a process of deliberation is implemented in functioning systems in which different parties discuss with one another on equal footing (Berkes 2007). The prerequisite for this is that all groups of actors in the network are really connected with one another (Pomeroy, Katon, and Harkes 1998).

Other authors point out that co-management regimes are particularly capable of balancing risks, using relationships productively, allocating tasks efficiently, and thus keeping transaction costs low (Carlsson and Berkes 2005). In addition, there is a higher sensitivity to local social, economic, and ecological conditions and limits, a higher sense of responsibility through clear rights, improved management through the use of local knowledge, and, finally, better compliance with the rules through local monitoring (Gutiérrez, Hilborn, and Defeo 2011).

Adaptive co-management is a relatively new field of research (Plummer and Armitage 2007). Some systematization research, such as a more comprehensive evaluation of case studies (Evans, Cherrett, and Pemsl 2011; but see Gutiérrez, Hilborn, and Defeo 2011), an indicator system, and a framework model are still lacking (Plummer and Armitage 2007). However, it is clear that the integrative character of this research, which focuses on feedback processes, adaptability, and community learning, represents a further step toward the analysis of complex SES and is therefore related to the present approach (Berkes, Colding, and Folke 2003). For example, it is repeatedly emphasized that various aspects of management, such as technical knowledge, are important but only a small part of the image alongside, for example, traditional knowledge (Armitage et al. 2009). This analysis agrees with this criticism and tries to present an overall view of SES complexities as well. For this purpose, twenty-four success factors are operationalized by several hundred variables in order to preserve as complex a representation as possible.

A strong connection to resilience research exists with the concept of learning and adaptability, which leads to an increased ability to deal with external shocks. Another important connection to SES research is the common emphasis on participation and rights of users involved. Many authors from both fields emphasize the difference between rules that are valid in formal terms (de jure) but are not actually implemented (de facto).

For the present work, besides the theoretical background, it is of importance that some of the papers from this area identify success factors. Success factors in co-management are discussed in several studies and analyzed in part quantitatively (Armitage et al. 2009; Evans, Cherrett, and Pemsl 2011; Gutiérrez, Hilborn, and Defeo 2011; Pomeroy, Katon, and Harkes 1998; Cinner et al. 2012). The last three syntheses are included in Section 2.5 and discussed in detail. Overall, the implementation of co-management is assessed as positive in many social, environmental, and economic indicators (e.g., Evans, Cherrett, and Pemsl 2011; Gutiérrez, Hilborn, and Defeo 2011; Cinner et al. 2012).

To conclude: both general analytical framework and conceptual apparatus are defined (Section 2.4.2). The global importance of natural resource management has been demonstrated and briefly characterized in terms of its different forms of management (see Sections 1.1 and 2.4.1). The status of research on social dilemma situations in jointly managed resources has been presented (Section 2.3), and SES have been described as complex systems with many cross-disciplinary factors (Sections 2.1.2 and 2.4.2). With this preliminary work, it is now possible to analyze in greater detail the field of research that deals directly with success factors in SES.

2.5 Potential success factors for sustainable management of social-ecological systems

The approaches from biological cooperation research and Experimental Economics considered so far both provide valuable indications of what promotes and/or prevents cooperation. They constitute a bridge for some of the fundamental behaviors of human beings; their effectiveness is explained through evolutionary mechanisms. However, they suffer from their generality and their lack of ecological validity with regard to the initial question as to which factors for sustainable management in SES are important for success. Although they represent the behavioral theoretical background, *specific* predictions for the success of collectively acting local institutions are not possible.

Therefore, the hitherto fairly broad focus on institutions in SES will be narrowed down. The focus is now on cooperation within SES, especially for CPR. I will take up the indications given previously again and again in the following, as they point out basic limitations, mechanisms of action, and potentially important factors.

Early analyses often looked for an underlying success factor. This has failed. On the one hand, there is not the one success factor that is common to all since institutions are too diverse. On the other hand, system complexity is high. Therefore, just one factor may very likely be not enough. Furthermore, almost all "panaceas," that is, the transfer of a successful approach to other institutions, have failed in the past (Ostrom 2007; Meinzen-Dick 2007). Particularly in policy advise and in the work of NGOs (non-governmental organizations), too often too simple plans, models, and instructions for action are still being worked with. They do not take sufficient account of SES complexity (Ostrom et al. 2011; Ostrom 1992a). Many development cooperation projects therefore often have disappointing end result (Ostrom and Cox 2010; Ostrom et al. 2011).

2.5.1 Design principles

Nevertheless, there is probably a set of design principles that make success very likely under various circumstances. Here Elinor Ostrom has done groundbreaking work (Ostrom 1990). Her definition of such construction principles is:

> By "design principle" I mean an essential element or condition that helps to account for the success of these institutions in sustaining the CPRs and gaining the compliance of generation after generation of appropriators to the rules in use. (Ostrom 1990, p. 90)

In *Governing the Commons*, she draws on an empirical analysis of fourteen case studies to produce a list of design principles that she considers fundamental to success. The eight factors cited by her can be considered as a basic framework for more comprehensive and later syntheses.

1. Clearly defined boundaries
Individuals or households who have rights to withdraw resource units from the CPR must be clearly defined, as must the boundaries of the CPR itself.

2. Congruence between appropriation and provision rules and local conditions
Appropriation rules restricting time, place, technology, and/or quantity of resource units are related to local conditions and to provision rules requiring labor, material, and/or money.

3. Collective-choice arrangements
Most individuals affected by the operational rules can participate in modifying the operational rules.

4. Monitoring
Monitors, who actively audit CPR conditions and appropriator behavior, are accountable to the appropriators or are the appropriators.

5. Graduated sanctions
Appropriators who violate operational rules are likely to be assessed graduated sanctions (depending on the seriousness and context of the offense) by other appropriators, by officials accountable to these appropriators, or by both.

6. Conflict-resolution mechanisms
Appropriators and their officials have rapid access to low-cost local arenas to resolve conflicts among appropriators or between appropriators and officials.

7. Minimal recognition of rights to organize
The rights of appropriators to devise their own institutions are not challenged by external governmental authorities.

8. Nested enterprises (only for CPR that are parts of larger systems)
Appropriation, provision, monitoring, enforcement, conflict resolution, and governance activities are organized in multiple layers of nested enterprises (Ostrom 1990, p. 90).

Are these factors the most important success factors? Are they cross-sectoral (fisheries, forest management, irrigation)? How far are they generalizable? Are there any further equally important factors missing in this list?

These and other questions have been asked. The answers are presented in the remainder of this section. The fourteen cases studied by Ostrom prove these factors to be significant, since they can be classified as clear successes or failures, depending on whether they exhibit these characteristics or not (Ostrom 1990, p. 180). A further case study confirms these eight factors (Nilsson 2001). As far as I know, there are no further comparative studies, although a database of eighty-six cases is available (CPR database, see Section 3.1).

Twenty years later, an overview study comes to the conclusion that these design principles have been empirically and theoretically examined by ninety-one studies (Cox, Arnold, and Villamayor Tomas 2010). For all factors, there is a clearly positive ratio of studies that confirm them as relevant to those who do not. This ratio ranges from 2.0 for graduated penalties (effect size 0.262) to 11.7 for monitoring (effect size 0.792). The conclusion is that these design principles are empirically well confirmed and relevant for the success of CPR systems. Since both the original case studies (Ostrom 1990) and the ninety-one case studies since then (Cox, Arnold, and Villamayor Tomas 2010) cover different sectors, pending a more detailed analysis (see Sections 5.2, 5.3, 5.4, 5.5), it must be assumed that these success factors are general and abstract enough to apply to SES across all sectors.

However, the question of whether all relevant factors are included must be answered negatively.

One of the first attempts to extend this synthesis of factors comes from Ostrom herself and identifies eleven factors (Ostrom 1992b). Furthermore, many authors suggest or differentiate further factors within the original eight factors (Cox, Arnold, and Villamayor Tomas 2010). Thus, these factors represent *general* environmental conditions for individuals. However, a finer modeling of individual differences remains impossible, since both a theoretical model and the necessary data are missing in this resolution.

2.5.2 Overview about success factor syntheses

This section provides an overview of attempts to identify success factors or system attributes (Cinner et al. 2012), or key principles (Gruber 2008), or conditions for sustainability (Agrawal 2001). The syntheses are presented and briefly discussed. In the discussion, I will focus on the most important aspects. The respective discussion makes no claim to completeness.

This comparison of existing syntheses pursues several objectives. Such an overview does not yet exist, so it can be used for further systematization. It is also the basis for an overall synthesis as I conduct it (Sections 4.2 and 5.1). In addition, it clarifies which meta-criteria (scale, scope, categories) are important, since each synthesis has to make compromises, for example regarding the number of concepts.

2.5.2.1 Synthesis 1 (synthesis of success factors)

I start with an attempt to identify a joint synthesis of several authors (Agrawal 2001; cf. also Agrawal 2002). Agrawal obtains an extensive synthesis of factors from Ostrom (1990), Baland and Platteau (1996), and Wade (1994) as well as from his own additions. Success is defined here as the stability of institutions. However, this is not operationalized. Since the factors of the investigated authors overlap strongly, it is possible to speak in large parts of a consensus. The success factors are divided into four categories with their respective interfaces (Agrawal 2001): resource, group, institutional arrangements, and external environments.

1. Resource
   - Small size (i.e. a small size of a system is more conducive to success)
   - Well-defined boundaries
   - Low levels of mobility
   - Possibilities of storage of benefits from the resource
   - Predictability
2. Group
   - Small size
   - Clearly defined boundaries
   - Shared norms
   - Past successful experiences—social capital
   - Appropriate leadership—young, familiar with changing external environments, connected to local traditional elite
   - Interdependence among group members
   - Heterogeneity of endowments, homogeneity of identities and interests
   - Low levels of poverty
3. Resource and groups
   - Overlap between user group residential location and resource location
   - High levels of dependence by group members on resource system
   - Fairness in allocation of benefits from common resources
   - Low levels of user demand
   - Gradual change in levels of demand
4. Institutions and rules
   - Rules are simple and easy to understand
   - Locally devised access and management rules
   - Ease in enforcement of rules
   - Graduated sanctions
   - Availability of low-cost adjudication
   - Accountability of monitors and other officials to users

5. Resource, institutions, and rules
   - match restrictions on harvests to regeneration of resources
6. External environment
   - Low cost exclusion technology
   - Time for adaptation to new technologies related to the commons
   - Low levels of articulation with external markets
   - Gradual change in articulation with external markets
   - Central government does not undermine local authority
   - Supportive external sanctioning institutions
   - Appropriate levels of external aid to compensate local users for conservation activities
   - Nested levels of appropriation, provision, enforcement, governance

As a supplementary point, the establishment of a community money fund can be mentioned, which can alleviate social hardship in local communities that act collectively, for example in the event of crop failure. Such a fund may contribute to greater justice as well (Wade 1992).

In this compilation, particular attention must be paid to the additions of the role of the markets. Success factors affecting external markets are rarely found in the original syntheses. Another special feature of this compilation are added links between categories. This draws attention to the difficulty of clearly separating facts into categories. Often a factor can be assigned to two or more categories. For example, the adjustment of the appropriation level affects both the rule system and the resource. Virtually all factors are included in the SES framework model (Synthesis 10, see Section 2.5.2.10).

2.5.2.2 Synthesis 2 (fisheries in Asia)

A synthesis of twenty-five research projects after five years of practical project research proves the following factors to be important for successful participatory management in Asian fishing cooperatives (Pomeroy, Katon, and Harkes 1998; Pomeroy, Katon, and Harkes 2001). They are also based on Ostrom's synthesis of factors.

The following factors have a *high* importance for success:

1. Clearly defined limits of the resource
2. Clearly defined limits of the user group
3. Positive cost-benefit balance
4. Joint institutions (participation of users in rule changes)
5. Easily understandable rules that are actually implemented; monitoring and enforcement by all parties involved
6. Willingness of users to participate at local level in the management of money, labor, and time

The following factors have a *medium to high* importance for success:

7. Group cohesion (consisting of homogeneity of the group, overlap of the group's place of residence, and the resource)
8. Existing organizations and experience with co-management
9. Local rights accepted by the state government, that is, laws exist which allow self-organization
10. Local groups authorized by the state, which take over the administrative and organizational tasks for the co-management between government and community (via a mediating committee)

The research group adds another twenty-eight self-developed success factors, some of which overlap or are repeated. Here is a shortened, non-redundant list:

11. Mechanisms for conflict resolution exist
12. Locally based leadership group or personality
13. Effective enforcement of penalties (in the organization and through government legislation)
14. Property rights are clear and secure
15. Support through national and local policies (recognition and empowerment, existing legislation)
16. Flexibility and adaptability to new problems
17. Fit with existing and traditional social and cultural institutions and structures of the community
18. Appropriately sized scale (for all areas)
19. Building knowledge and skills for collective action
20. Existence of incentive structures that work individually
21. Problems managing the resource can be detected
22. Adequate preparation time (especially of social values and moral standards)
23. Objectives and problems are clear
24. Effective communication
25. Political and social stability
26. Existence of financial resources and a budget plan
27. Involvement of external stakeholders and consultants
28. Trust between parties (e.g., government and community)
29. Networking of community organizations
30. Legitimate organizations (with clear admission rules)
31. Working coordination between partners (common interests, associations, contracts, mediating body)
32. Partners feel they are involved in the process

It becomes clear in this compilation that a structure, for example with a framework, is missing. For example, "effective communication" and "appropriate scale in size" are criteria that can be applied to different actors. "Political and social stability," on the other hand, characterizes the environment in which the problem is embedded, but does not belong to the institutions. In addition, the aspect "benefit exceeds cost" describes a meta-criterion that must always be given but cannot be measured in practice. The high importance of other organizations and groups is striking, which probably results from the practical experience of the authors, who are active in many consulting projects.

2.5.2.3 Synthesis 3 (meta-analysis forestry worldwide)

One of the few existing meta-studies on success factors in SES identifies forty-three success factors from sixty-nine case studies on community forest management (Pagdee, Kim, and Daugherty 2006). This study explicitly points to the problem of defining success (see also Section 4.2.6) and measures success in three dimensions, namely ecological, social, and economic success. At the same time, the study notes that most case studies examine only one dimension. The heterogeneity of the studies makes a classification of factors that would go beyond coarse dichotomies impossible (large / small, existent / non-existent). The success factors of this meta-analysis are as follows:

1. Regulation of ownership relationships
   - Security of ownership of the resource
   - Clear ownership of the resource
   - Clear limits of the resource
   - Designated areas for special use
   - Congruence of the biophysical properties of the community and the resource and social boundaries
   - Rules for the use of the products
2. Institutions
   - Effective enforcement of rules
   - Monitoring
   - Penalties
   - Participants have experience with self-government
   - Strong leadership
   - Local organization has financial resources and manpower
3. Incentives and interests
   - Resource has value (sets self-government in motion)
   - Costs of self-government and institutional change
   - Expectation that the benefit will benefit those involved
   - Dependency on resource

- Common interests (sets self-government in motion)
4. Outside support
   - Willingness of the government to implement self-government
   - Support from financial resources and labor (national and international governments, non-governmental organizations, individuals)
   - Technical support from external experts
5. Biophysical characteristics
   - Size of the resource
   - Geographical location (access)
   - Biodiversity of species
   - Status of the resource
   - Status of the resource (future trend)
   - Predictability of the appropriation level of the units
6. Characteristics of the community
   - Size of the community
   - Location (proximity to resource)
   - Rising population growth
   - Increasing inflow and outflow
   - Conflicts between local community and external parties
   - Socio-cultural diversity (heterogeneity)
   - Community economic situation
   - Experience of the community with cooperative work
   - Traditional practices
7. Share of community that participates in self-government
   - Higher proportion of community members participating in management projects with other partners leads to better results
8. Extent of decentralization
   - Level of recognition (legal or informal recognition; self-administration rights, no rights)
   - Clear process rules to carry out local controls
   - Management is delegated to local community (local responsibility)
   - Financial budget transfer to local community (local authority)
9. Influence of technology and markets
   - Changes in technology
   - Greater demand for products from the resource and higher economic value
   - Introduction of infrastructure
   - Instability and fluctuation of market conditions

This compilation is also astonishingly congruent with the previous lists: Important components are the characteristics of the resource system, the attributes of the community, the regulatory systems (above all secure property rights), the connection to the outside world (connection to markets and networking with other actors), as well as the participation of its members, which in turn depends strongly on the incentive structures. The most important individual factors are "overlap of the natural limits of the resource with socio-economic limits" and "effective implementation of rules." This synthesis of success factors is an important support for my integrated synthesis, since it is comprehensive, balanced, and logically consistent in the level selection.

2.5.2.4 Synthesis 4 (small-scale forest management in Germany)

Interestingly, the factors for overcoming an anticommons situation, that is, a strong privatization, are quite similar. One example is German forests (Schurr 2006). Here, relevant factors are:

1. Creation of rights (co-management, appropriation, access, management, exclusion, and disposal)
2. Clear boundaries
3. Institutions that set the general rule system
4. In the foundation phase: monetary support of the right to self-organize is important, since transaction costs for overcoming the anticommons situation are otherwise prohibitively high
5. Providing information, such as knowledge of ownership or the status of the CPR
6. Confidence building through cooperation: this reduces the risks for the collective organization
7. The withdrawal of the state provides incentives for self-organization, which must be coordinated by key persons

This study focuses on initial obstacles to self-organization. Self-organization is reliably prevented by an opaque legal situation and a small amount of information about group and resource.

2.5.2.5 Synthesis 5 (meta-analysis of local communities worldwide)

The next, detailed synthesis of factors emphasizes the rule systems (Gruber 2008). It was developed for the World Bank from case studies on the joint management of natural resources and aims to provide a principles matrix for theorists and field studies alike. Success is defined as the sustainable and effective management of shared natural resources:

1. Principle A: Public participation and mobilization
   • Effective public participation is integral to all forms of CBNRM and other community-based environmental initiatives

- Public participation will directly impact public trust, confidence, and legitimization
- Seek diversity of stakeholders, including citizens, NGOs, local and regional governments, private sector, and those with programmatic, operational, scientific, and legal knowledge
- Provide for participation of stakeholders at all stages: information gathering, consultation, visioning and goal setting, decision making, initiating action, participating in projects, and evaluation

2. Principle B: Social capital and collaborative partnerships
   - Networks and partnerships are integral to building social capital and serve as a catalyst to finding innovative strategies and solutions
   - Collaborative partnerships are key to leveraging resources and supporting implementation
   - Stakeholder trainings, workshops, and other collaborative learning opportunities can build social capital and commitment
   - Seek agreement among key environmental NGOs, governments, and the private sector to work collaboratively and to share resource and responsibilities
   - Ownership by community members and other stakeholders enhances design, implementation, and operation; supports cohesion; and encourages long-term commitment

3. Principle C: Resources and equity
   - Environmental justice is a social imperative that includes recognizing local values
   - Seek to improve (or minimize negative effects on) the local economy
   - Recognize need for linkages between conservation and local economy based on equity, local needs, and financial and environmental sustainability
   - Seek equitable and fair distribution of local benefits, potentially including compensation for protecting natural resources
   - Regulated access to natural resources and graduated sanctions can help ensure equity

4. Principle D: Communication and information dissemination
   - Well-designed communication systems provide information sharing that support multiple social networks and raises levels of knowledge and awareness
   - Linkages are provided between different information and knowledge systems to support learning, decision making, and change
   - Effective communication supports openness and transparency
   - Promote information sharing between experts and non-experts though multiple approaches, including seminars and workshops; printed, electronic, and mass media; and projects
   - Explicitly state expectations and limits

5. Principle E: Research and information development
   - There is a common information base that is accessible and useful
   - Decisions should be based on a broad but systematic body of information
   - Integrated information includes technical, scientific, social, quality-of-life, economic, and other forms of local knowledge, including indigenous experiential knowledge
   - Economic evaluation of environmental assets is a valuable information base
   - Ongoing research is necessary to improve on existing solutions, including a role for community members in collection of scientific information

6. Principle F: Devolution and empowerment
   - True sharing of power and responsibility (devolution of authority and responsibility) between government authorities, community groups, and the wider community with enhanced local decision making improves outcomes
   - Most individuals affected by environmental rules and regulations, including those who are often marginalized, should be included or represented in the group who make or modify the rules
   - There are nested, multiple layers of governments and enterprises related to role and activities of decision making, appropriation, monitoring, enforcement, conflict resolution, and governance
   - Devolution of control and decision making significantly changes the relationship between central governments and rural and regional areas and, if done effectively, can engage and build commitment of local community members
   - Establishing clear rules, procedures, and regulations can empower the local community

7. Principle G: Public trust and legitimacy
   - Work must be viewed by community as legitimate to build community trust
   - Local leaders are integral to efforts in establishing trust and credibility
   - Support by local elected officials will build trust and legitimacy
   - Participatory approaches to problem solving and decision making are critical to building legitimacy
   - Transparency in activities, including decision making, supports the building of trust

8. Principle H: Monitoring, feedback, and accountability
   - Tight feedback loops are supported by openness, transparency, monitoring, mutual accountability, collaboration, and power sharing between the stakeholders and partners
   - Effective feedback systems, including feedback from social networks, allow for opportunities to learn from mistakes, uncertainty, and crises

- Local appointed or elected representatives of communities must themselves be accountable to their constituents if community-based conservation is to be responsive to the community
- The performance of those who make decisions should be periodically reviewed by those that are affected by the decisions
- The social and technical capacity for monitoring, evaluating, responding, and enforcement is necessary for effective and dynamic systems

9. Principle I: Adaptive leadership and co-management
   - A robust social-ecologic organization is designed and supported to be a learning organization that supports adaptive capacity
   - A learning organization and an optimum management system is resilient to perturbation, with an ability to cope with external shocks and rapid change
   - Adaptive co-management and adaptive leadership are dynamic and focused on processes rather than static structures
   - Adaptive co-management approaches include roles for local government, local community members, NGOs, and private institutions and decision making inclusive of people affected by and knowledgeable of the issues
   - An effective co-management approach engages, trains, and mobilizes community members in the work of the organization

10. Principle J: Participatory decision making
    - Effective participatory problem solving and decision making is enabled by a well-structured and well-facilitated dialogue involving scientists, policy makers, resource users, practitioners, and community members
    - Decision making is informed by analysis of key information about environmental and human-environmental systems, including life aspirations of local people
    - It is vital to create a shared holistic vision/plan that anticipates probable environmental, social, and economic outcomes
    - The policy creation process should include a wide range of key expert and nonexpert constituency and community groups "at the table"
    - Participatory problem solving should provide opportunities for the sharing of knowledge and collaborative learning about SES

11. Principle K: Enabling environment: Optimal preconditions or early conditions
    - Community has a homogeneous social structure, common interests, shared norms and a local social structure in which divisions are not too serious or disruptive of cooperation
    - There are clearly defined boundaries of the resource system
    - The public is unsatisfied with the status quo but is not feeling hopeless

- Citizens and stakeholders are willing to participate because they have a high sense of community and/or dependency on the local natural resourc
- There is adequate support and investment of financial and other resources to support transitional costs
12. Principle L: Conflict resolution and cooperation
    - Difficult realities and conflicts are inherent in community-based SES
    - Plan for and develop capacity and strategies for conflict management and resolution at the time of initiation of a community-based social-ecologic initiative
    - Recognize the central role of institutions outside of the community-based organization in mediation of environment-society conflicts
    - Work to transcend organizational rivalry and competition between organizations or stakeholder groups
    - Design participatory decision-making processes that promote dialogue and reduce factionalism

This meta-study extracts the twelve factors from a randomized sample of community-based natural resource management projects from twenty-four studies in twenty-three countries. It attaches great importance to the transfer of responsibility to local actors, transparency of regulatory processes, existence of social networks, and involvement of all participants at different levels in an open, information-based, joint decision-making process. In contrast, biophysical factors such as the properties of the resource system and resource units receive little attention. The level of abstractness of the success factors of these syntheses is roughly on the same level so far. The factors are more abstract and more general than individual variables. Other studies, like the following synthesis, do not take this step and work on the degree of resolution of variables.

2.5.2.6 Synthesis 6 (irrigation systems in India)

The following list of positive factors for the condition of irrigation systems in a study of eighty-seven Indian villages shows such a lower degree of resolution (Shiferaw, Kebede, and Reddy 2008):

1. High rainfall (causes a low appropriation)
2. Distance to the nearest market
3. Number of castes (positive effect of heterogeneity in the group)
4. Number of telephones (as an indicator of market integration)
5. Education level
6. Preferred employment of women as workers
7. Reading out the minutes of the last meeting of the institution (most important factor in this study)

Negative correlations with the condition of the irrigation system exist with:

1. The larger geographical extent of the villages
2. The number of households
3. The distance to the local administration
4. The higher proportion of households with electricity
5. A higher proportion of overexploited land

A disadvantage of this study for a theoretical-conceptual study such as the present one is the lack of generalizability, advantageous the more precise degree of resolution compared to the other compilations. Thus, it is possible to divide the size (number of households and size of the village) into two components—both with negative influence—one with influence on the institutional capacity for collective action, the other with a view to mobilizing members. General factors are not always present.

2.5.2.7 Synthesis 7 (meta-analysis fisheries, worldwide)

A meta-analysis of global fisheries identifies nineteen success factors (Gutiérrez, Hilborn, and Defeo 2011). All fishing collectives investigated are operated jointly by the state and local actors (co-management). As in synthesis 3 (Section 5.2.3), success is divided into ecological, economic, and social success and operationalized accordingly. The success factors are arranged according to decreasing importance:

1. Existence of a leadership personality or group (leadership)
2. Individual or community individual quotas
3. Cohesion and trust of the group (fishers)
4. Designation of areas where fishing is prohibited or the presence of no-take or MPAs (marine protected areas)
5. Enforcement of the rules by the community (self-enforcement)
6. Long-term management plans (long-term management policy)
7. Monitoring, control, and surveillance
8. Influence in local markets
9. Territorial user rights for fishing (TURF)
10. Low or no mobility of resource units (sedentary/low mobility)
11. Spatially explicit management (identification of corresponding areas)
12. Defined boundaries
13. Fishing quotas for the community (global catch quotas)
14. Central government support (local)
15. Seeding or restocking
16. Scientific advice
17. Minimum sizes for catches

18. Co-management in national law
19. Tradition

With regard to fisheries, this is probably the largest study in terms of numbers, with 130 fisheries all under co-management (Gutiérrez, Hilborn, and Defeo 2011). While it analyses mostly smaller communities, there is another study concentrating the other end of the spectrum—very large marine protected areas (Ban et al. 2017). The seven most important factors are the relevance of a functional group (leadership, social cohesion), adaptability (individual quotas), stability (long-term management plans), and actual enforcement of the rules (monitoring, self-enforcement) as well as regeneration possibilities of the resource (marine protected areas). This study also clearly shows that success factors must include both bio-physical conditions as well as a group of actors and control systems (GS)—only in combination do they become effective. It also demonstrates that there are certain limits to what can be achieved—factors gain importance under different conditions (such as a higher level of development in the country, as indicated by the HDI, the human development index) or another type of fishing. Nevertheless, the most important factors remain constant in most combinations, albeit in a different order.

2.5.2.8 Synthesis 8 (fisheries in Kenya, Tanzania, Madagascar, Indonesia, and Papua-New Guinea)

Another study on co-management compares small fishing collectives fishing at coral reefs across countries (Cinner et al. 2012). The relevant success factors are:

1. Distance to market
2. Population size
3. Wealth
4. Years of education
5. Age of organization
6. Trust in leaders
7. Trust in the community
8. Share of migrants
9. Frequency of participation in community events
10. Human agency's role in the state of the resource
11. Fishing as a primary marine livelihood
12. Occupational diversity per household
13. Access restrictions
14. Clear boundaries of resources
15. Participation (taking part in decisions)
16. Graduated sanctions

All forty-two systems examined are under co-management (Cinner et al. 2012). The success factors are assigned to the categories of the SES framework model (see Section 2.5.2.10, synthesis 10). Environmental success—measured in this study with a fish biomass indicator—is higher when markets are more difficult to reach, when the community is less dependent on the resource, when there is confidence in the leaders, and when there are fewer jobs per household (Cinner et al. 2012). The relevance of strong leadership in fisheries (Gutiérrez, Hilborn, and Defeo 2011) is confirmed here. The important role of institutions is also confirmed, for example, their influence on compliance. However, this study finds no direct connection between institutions and ecological success.

2.5.2.9 Synthesis 9 (nature conservation projects, worldwide)

This study covers 136 nature conservation projects worldwide that are not sector-specific (Brooks, Waylen, and Borgerhoff Mulder 2012). It is about success of local groups in sustainable management of natural resources (community-based conservation). The identified factors are:

1. Rights on a national level
2. GINI-index
3. Level of participation
4. Level of engagement
5. Protectionism
6. Equity
7. Capacity building
8. Social capital
9. Environmental education
10. Threat
11. Tenure
12. Charisma
13. Population size

A medium to high degree of participation, close integration with the local community, and the development of skills are important for ecological success. Economic success, on the other hand, is associated with owning or controlling the resource and building up capacity. A small population size (<5,000), effective local institutions that support nature conservation, and the development of skills are important for the success of the project in terms of conservation of nature (behavioral success). Medium or high participation opportunities, fair distribution, and social capital are important for local acceptance (attitudinal success).

2.5.2.10   Synthesis 10 (synthesis of success factors, social-ecological systems framework)

Perhaps the most influential set of factors is the SES framework (Partelow 2018). It has been widely used for many different SES (e.g., Blythe et al. 2017; Addison and Greiner 2016; Leslie et al. 2015). It divides thirty-three system attributes into four groups: resource, resource units, actors, and governance systems (Ostrom 2007). These four subsystems interact with each other and are set within a larger socio-economic and ecological environment (see Figure 2.6).

An important difference to design principles, for example, is that this list sees itself as a theoretical framework for the analysis of SES and not as a synthesis of success factors. Therefore, some factors from Ostrom's earlier work can also be found here in both a more general form and a neutral formulation as components of SES. This makes sense since some of these factors are not linear from negative to positive. For example, both too small and too large a number of actors is detrimental for cooperation. Too small a number often prevents the creation of a collective, whereas too large a number makes effective monitoring difficult.

However, all components of the four main categories can also be regarded as success factors, such as resource unit mobility. Mobility is, on the one hand, a description of an important attribute of resource units. For the analysis of SES, this is a useful distinguishing criterion between sectors, such as fish and trees. On the other hand, its character as a success factor quickly becomes clear: in most cases, high mobility is associated with difficulties during removal and increases costs (Schlager, Blomquist, and Tang 1994). The so-called second-tier variables are:

1. Resource (Resource systems, RS)
   - RS1 Sector (e.g., water, forests, pasture, fish)
   - RS2 Clarity of system boundaries
   - RS3 Size of resource system* (variables marked with * are decisive whether self-organization takes place)
   - RS4 Human-constructed facilities
   - RS5 Productivity of system*
   - RS6 Equilibrium properties
   - RS7 Predictability of system dynamics*
   - RS8 Storage characteristics
   - RS9 Location
2. Resource units (RU)
   - RU1 Resource unit mobility*
   - RU2 Growth or replacement rate
   - RU3 Interaction among resource units

- RU4 Economic value
- RU5 Number of units
- RU6 Distinctive markings
- RU7 Spatial and temporal distribution
3. Governance systems (GS)
   - GS1 Government organizations
   - GS2 Non-government organizations
   - GS3 Network structure
   - GS4 Property-rights systems
   - GS5 Operational rules
   - GS6 Collective-choice rules*
   - GS7 Constitutional rules
   - GS8 Monitoring and sanctioning processes
4. Users (U)
   - U1 Number of actors*
   - U2 Socioeconomic attributes of actors
   - U3 History of use
   - U4 Location
   - U5 Leadership/entrepreneurship*
   - U6 Norms/social capital*
   - U7 Knowledge of SES/mental models*
   - U8 Importance of resource to actors*
   - U9 Technology used

Another category are the activities (action situations) in such systems (interactions, I1–I8). They are descriptions of central processes in SES and produce results (outcomes [O]) together with framework conditions (social, economic, and political settings [S]), and related ecosystems (ECO):

1. Interactions → outcomes (I → O)
   - I1 Harvesting levels of diverse actors
   - I2 Information sharing among actors
   - I3 Deliberation processes
   - I4 Conflicts among actors
   - I5 Investment activities
   - I6 Lobbying activities
   - I7 Self-organizing activities
   - I8 Networking activities
2. Outcomes (O)
   - O1 Social performance measures, for example, efficiency, equity, accountability, sustainability

- O2 Ecological performance measures, for example, overharvested, resilience, bio-diversity, sustainability
- O3 Externalities to other SES

3. Related ecosystems (ECO)
   - ECO1 Climate patterns
   - ECO2 Pollution patterns
   - ECO3 Flows into and out of focal SES
4. Social, economic, and political settings (S)
   - S1 Economic development
   - S2 Demographic trends
   - S3 Political stability
   - S4 Government resource policies
   - S5 Market incentives
   - S6 Media organization

The SES framework model has proved to be extremely effective—many studies use it as a theoretical starting point for their analyses and consider partial aspects (e.g., Schlüter and Madrigal 2012; Baur, Liechti, and Binder 2014). It is under constant revision (e.g., Hinkel et al. 2015; McGinnis and Ostrom 2014). However, it has also been criticized to be too much actor-centered (Partelow 2015) and too unspecified for the ecological side of SES (Vogt et al. 2015).

2.5.3 Summary of syntheses for social-ecological systems

Several syntheses mention good leadership, opportunities for participation, clear boundaries and property rights, and interdependence with markets as factors for success. However, the ten factor syntheses do not show any success factors that consistently prove to be important for ecological success throughout several studies or meta-analyses, with the exception of perhaps a high degree of participation by the local community. This is also due to the different sectors, the different methodological approaches, and a different degree of resolution. Another important difference is the use of different concepts, which could explain the different results. Even if the same or similar concepts are used, the operationalization varies from study to study.

For our synthesis of these studies, that is, the twenty-four success factors on which our analysis is based, see Section 5.1.3.

It is therefore not surprising that no reliable conclusion can be drawn from the previously mentioned studies in regard to robust success factors, even though some studies consider many case studies or meta-analyses. Partly, this lack motivates the research presented here.

For this reason, Chapter 4: Methods and Chapter 5: Results analyze three large datasets whose data are collected uniformly; their concepts were developed by Elinor Ostrom, the leading theorist on SES and common-pool resource problems. They are analyzed with different statistical methods in order to validate the results so that the findings regarding the success factors should apply to a large number of different systems. First, the data and methods used are described in detail.

3

Data

There are only a few large databases that collect information about natural resource systems (Poteete, Janssen, and Ostrom 2010). Compared to a meta-analysis of individual studies, they have the advantage that all cases are subject to a uniform system. The present analysis is based on three such data sets of social-ecological systems (SES), which were developed at the Ostrom-Workshop in Bloomington. Immediate advantages are that the theoretical framework (Ostrom 2009) is identical for all databases (CPR—for common pool resources, NIIS—for Nepal irrigation institution study, and IFRI—for International Forestry Resources and Institutions) and that two databases (CPR, NIIS) have a practically identical structure. In addition, CPR and NIIS share great similarities with the third (IFRI, http://www.ifriresearch.net/) since the latter is a further development of them. The CPR database contains 86 case studies of irrigation systems and fisheries worldwide with information collected on about 600 independent variables. The NIIS database has information on 244 Nepalese irrigation systems with about 600 variables. Finally, the IFRI database contains 455 cases on forestry worldwide with about 1,000 variables. The actual number of cases is higher, since the unit of analysis is not one location, but a user group that extracts a resource unit from a resource system applying one rule system. Thus, the CPR database yields 122 cases, the NIIS 263, and IFRI 409.

These case study collections are unique as they combine detailed biophysical and social information—variables are, for example, about the rules in place, institutions, actor groups, and other governance information but also on size, location, technological features of the system, and its ecological condition.

In order to make this more concrete, a list of typical questions for each question type including answer and data type follows:

1. Yes/No-questions
Example <ADJOINFD>: "If labor force is organized in small teams, is each team formed by individuals who farm adjoining fields?"

2. Text (e.g., description)
Example <PARAGRPH>: "Please write an abstract of the document being screened."

Sustainable Governance of Natural Resources. Ulrich Frey, Oxford University Press (2020). © Oxford University Press.
DOI: 10.1093/oso/9780197502211.001.0001.

3. Number
Example <SURFAREA>: "Surface area of resource in square meters."

4. Likert-scale (up to eleven values, mostly five)
Example <BEGBLNC>: "For biological resources at the beginning and end of this period, the balance between the quantity of units withdrawn and the number of units available is"
(1) _____ Extreme shortage, (2) _____ Moderate shortage, (3) _____ Apparently balanced, (4) _____ Moderately abundant, (5) _____ Quite abundant

5. Yes/No-chains
Example <CITCOUNT, CLAN, CASTE, GENDER, etc.>: "Please check off requirements that individuals must meet before appropriating from this resource." YYNN

The analysis of any large amount of data (here: 794 codeable case studies, each with values for about 500 to 1,000 variables) requires a systematic, structured, and at least partially automated procedure, since reproducibility, independence of persons, and comparability must be guaranteed for each working step. These steps are presented in the following sections. These data are checked for correctness (Section 3.5.1), are selected (Section 3.5.2), cleaned and processed (Section 3.5.3), and then prepared for statistical analysis (Sections 3.5.4 and 3.5.5).

3.1 Common-pool resource database

The CPR database contains global case studies on both fisheries and irrigation systems. It was the first case study collection of the Ostrom Workshop in Political Theory and Policy Analysis (Indiana University). Cases were recorded until 1987. The database contains 13 tables with a total of 593 variables for 86 case studies yielding 125 actor-rule. Of these 125 cases, 122 are included in the analysis. The 3 excluded cases concern other resources—forest management (2) and seaweed (1).

The cases have been selected from published studies—only studies with sufficient information on the majority of the variables of interest were included.

3.2 Nepal irrigation institution study database

The NIIS database contains only Nepalese irrigation systems. It was also developed at the Ostrom Workshop in Bloomington. The data have been collected during several stays in Nepal (Lam 1998). Some scholarly works from the Ostrom workshop are based on these data and describe them in more detail (Lam 1998; Tang 1992; Ostrom 1992a).

One important piece of information concerns the form of management: 21 cases (8%) are managed by the government (AMIS, agency-managed irrigation systems), 208 cases (79%) by users (FMIS, farmer-managed irrigation systems), 28 cases (11%) are jointly managed (JMIS, joint management cf. Frey, Villamayor-Tomas, and Theesfeld 2016).

All 263 codeable cases are used for analysis. The data were partly collected in field studies from 1982 to 1997 (rapid appraisals) and partly extracted from published studies. The 566 variables are coded exactly as in the CPR database (e.g., Yes/No, etc.). Mostly, these are exactly the same variables, because the NIIS database represents a further development of the CPR and both databases analyze irrigation systems.

3.3 International forestry resources and institutions database

The IFRI database is an extension of the CPR and NIIS to another resource—forestry management. However, it represents a greater adjustment both in terms of resource and number of cases as well as structure of the database (Poteete, Janssen, and Ostrom 2010). The analyses on the IFRI data were done in close cooperation with Dr. Carl Salk, a forest ecologist who has been working for IFRI in the field and has published numerous scientific articles on IFRI.

The project start for IFRI was 1992, however, data will continue to be collected (status 2019). Further information and publications can be found on the web page (http://www.ifriresearch.net/). In contrast to CPR, the data for IFRI do not mainly consist of already published work, but are collected by thirteen research centers according to a defined protocol by an interdisciplinary working group (Poteete, Janssen, and Ostrom 2010). The centers are located in India, Thailand, Nepal, Ethiopia, Kenya, Tanzania, Uganda, Bolivia, Guatemala, Colombia, Mexico, and the US (Indiana, Michigan). There are a variety of publications about IFRI, which can be found on the IFRI website (http://www.ifriresearch. net/resources/publications-2/). The important ones for this analysis will be discussed in detail later in Section 5.1.5 (e.g., Gibson, Williams, and Ostrom 2005; van Laerhoven 2010; Agrawal and Chhatre 2006).

The version used is the one prepared for the IASC conference in January 2011 in Hyderabad, India, by the Vincent and Elinor Ostrom Workshop for Political Theory and Policy Analysis. In total, the database contains 878 variables. Of these 878 variables, 208 are empty. This database also overlaps with the CPR and the NIIS database regarding the variables (see Section 3.4). The variables are also available in the form described for the CPR and NIIS database (e.g., Yes/No-questions), but there are proportionally more text variables. Of the 455 cases, 409 cases (90%) are included in the analysis. The remaining 46 forests are excluded since either wood products are not used, or there are no data, or the forest is intended for pure recreation. For further details, see Salk, Frey, and Rusch (2014).

3.4 Comparability of databases

The goal of the analysis is a quantitative model that is as precise as possible. The more case studies that flow into this model, the wider the basis, the more general statements can be made, and the more precise the underlying patterns can be identified by the neural networks. Therefore, it is important to make the individual databases (see Sections 3.1, 3.2, and 3.3) comparable.

Comparability is favored by the historical development of the databases, since they all have been developed by the same research group around Elinor Ostrom. The first database, CPR, established the theoretical basics, such as framework, concepts, and definitions of variables. The NIIS and IFRI databases developed and refined these analyses, since, according to Elinor Ostrom (verbal communication), NIIS and IFRI are each a further development of the CPR adapted to the respective resource.

Hence, the overlap of variables is considerable: Of the 593 variables used in the CPR, 319 are also found in the NIIS (54%). For IFRI, this comparison is no longer feasible because of another naming system. A rough estimate is around 20%. This overlapping is positive, as it increases comparability.

However, variable selection may be one-sided, thus overemphasizing social aspects and underemphasizing others (Vogt et al. 2015; Epstein et al. 2013). A potential systematic bias is mediated by the fact that variables have been created over many years, with different experts and on the basis of a meta-analysis. Furthermore, an empirical validation shows that the SES framework, which is based on these data, is comprehensive: Of 32 independent studies, which in turn investigate 1,749 cases and use a total of 656 variables, only 7% of the variables studied cannot be included in the SES framework model (Frey, unpublished data; Partelow 2018).

Comparability for the ecological success is also extremely important. To ensure this, the defined success factors remain the same across all databases. Each success factor is made up of indicators defined *before* the analysis. The variables that are different for each database are assigned to this fixed framework. This means that the individual content (type, number, and quality of individual variables) of the success factors may be different, but the factors themselves can be compared with each other.

To be able to compare across databases, the selected unit of analysis has to be the same. Here, it is a user group that extracts a resource unit from a resource system with a rule system. For example, a group of users who extract wood as well as mushrooms from a forest is considered to be two cases because wood and mushrooms represent two resource units—one might be harvested in a sustainable way, the other may not.

Cases are also split into separate cases if a user group makes fundamental changes to the rules-in-use. One would represent resource extraction with the first rule set and the second case would be extraction with the changed rule set. This separation is necessary because the rule system contributes decisively to the ecological success. For example, a first set of rules could lead to over-exploitation, while a second set of rules promotes sustainable management. The same consideration applies to the respective user group harvesting. Thus, within the same resource system, a user group can behave sustainably if the same unit is extracted (for example, wood in one-and-the-same forest), while another group of users does not (Ostrom 1990; Lam 1998). This choice of the analysis unit thus follows the database structure.

Finally, it has been doubted whether different resource types are comparable at all (Ostrom, Janssen, and Anderies 2007; Agrawal 2001). For this reason, first of all, a sector-specific model is developed, for example, a model for irrigation using the NIIS. Only then will all the encoded cases of all databases be integrated into an overall model. After that, the model quality of the sector-specific models can be compared with that of the overall model. This is done by predicting the success of the same test cases (e.g., irrigation systems in the NIIS) both in the overall and the NIIS model. If the overall model has more explanatory power (is more accurate), the possibility of comparing different sectors can be deduced post hoc as well as the relatively sector-independent nature of the success factors. If this is not the case, and the overall model scores worse than the individual model despite the higher number of cases, it either follows that the comparability is not given or that the success patterns are sector-specific and cannot be generalized. This is discussed in detail in Chapter 5: Results.

3.5 Data preparation

The preparation of data is a lengthy process involving many individual steps due to the complexity as well as the sheer amount of data. This process took three years and is described in the next section (Section 3.5.1). The specified steps are arranged chronologically, that is, they build upon one another in this order. The flowchart in Figure 3.1 shows this process in a simplified way.

The following sections (Sections 3.5.1–3.5.5) explain why each of these steps was necessary.

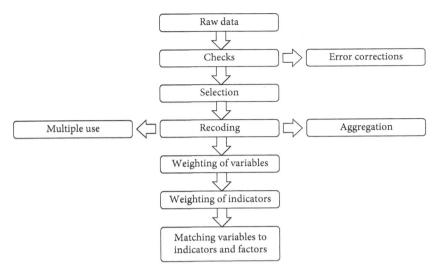

Figure 3.1 Flow chart of data processing

3.5.1 Check of raw data

Due to complexity and size, large amounts of data always entail the risk of errors. For example, it is impossible to visually check all data. The following sections (Sections 3.5.1–3.5.5) therefore discuss possible sources of error, precautions, and checks. The numerous recoding operations are then described in detail to ensure reproducibility.

3.5.1.1 Check for data correctness—step 1: Data collection
Some of the data on paper was collected years (IFRI) or decades (CPR and NIIS) ago. Therefore, only indirect checks on correct data collection are possible.

In the case of the CPR database there is a table (screener) which documents who has classified the underlying studies as suitable (screener), who has recoded them (coder), who has checked (reviewer), who has entered the recoding (coder entry), who has checked this again (coder check), and who has corrected it if necessary (coder correction). In addition to this multi-person test system (including Elinor Ostrom, who was responsible for content checks among other things), I randomly compared the recoded values in the database with the respective references, effectively doing a literature review. All these tests result in consistent and justified assessments throughout.

Most of the NIIS data has been collected directly in the field. Two things speak for the high quality. First, at least three dissertations have been published by persons who have worked with this data set both in the field and theoretically (Lam

1998; Tang 1992; Shivakoti and Ostrom 2002) as well as many articles (e.g., Joshi 2000; Lam 1996). Second, after the first data collection, missing data was *specifically* supplemented by completing systems already recorded in a second wave (Lam 1996).

The IFRI database is based on decades of experience with CPR and NIIS. Data structure, collection procedures, etc. can therefore be described as mature. In addition, many quality assurance measures were designed into it (Poteete, Janssen, and Ostrom 2010). For example, a special program has been developed for data entry which visually structures the content of the entry and checks it for impossible or incorrect entries.

Moreover, the data of individual research centers is checked at a central location (Michigan) for each version of the database that is released to the IFRI network. The quality of the data can therefore generally be regarded as high (for two exceptions and their correction see Salk, Frey, and Rusch 2014). Data collection via field observations is also consistently high, since a uniform protocol is used and a semi-annual course is required which conveys concepts and methodology (http://sitemaker.umich.edu/ifri/training) to become member of an IFRI team. In addition, data has always been recorded by an interdisciplinary team. The social science expert determines the social data, the biologist the biological data of the forest, etc. Particular attention is paid to ensuring that a forest expert and an expert for botanical names of this region are on the respective team.

3.5.1.2 Verification of the correctness of the data—step 2: Data entry

The original data for CPR, NIIS, and IFRI are recorded in paper form and are available at the Vincent and Elinor Ostrom Workshop for Political Theory and Policy Analysis. However, they are no longer fully available in CPR and NIIS. Despite extensive on-site searches in and enquiries at the respective libraries and archives, the missing documents could no longer be found. However, the existing data have been sampled for CPR and NIIS. No errors were found for data transmission from paper form questionnaires into the databases.

3.5.1.3 Verification of the correctness of the data—step 3: Databases

The original database format is Rbase for CPR and NIIS and Microsoft Access for IFRI. The first check was to re-export each table from Rbase and compare it with tables in an existing Access version. This resulted in about 2–5 errors per table for NIIS. These errors were corrected by either consulting the database expert, who also entered the data or by referring back to the original paper forms (e.g., for IDs), but above all by using internal logic (e.g., duplications, incorrect assignments).

Another step was to check internal consistency: The databases CPR, NIIS, and IFRI are available as linked tables. These are called relational databases. However, the links are without referential integrity. Changes to a table are therefore not passed on to tables linked to it. For this reason, data inconsistencies and errors can occur. Since it is possible to add referential integrity post hoc to a database, this can serve as a check. If there are no inconsistencies, as in IFRI, referential integrity can be added easily. This was not the case in CPR and NIIS. Thus, all the fixes described in the previous paragraph had to be applied before referential integrity could be added to CPR and NIIS, thus excluding any inconsistencies between tables. After this step, all three databases were without inconsistencies.

In addition, contents of the overview table were also checked against the individual tables. Here, the check was: Does the existing overview table, which combines the individual tables, really match the individual tables?

Here, too, a few corrections (NIIS: <5 out of 263 cases) were necessary, mostly spelling mistakes in the IDs of cases. These unique primary keys could be used to add a table (agriculture) that was not in the summary table. In another table (orginven), systematic duplications were removed. CPR's export from Rbase to Access was error-free. Moreover, the examination of the tables showed practically no errors. The few incorrect assignments could be resolved quickly by manual corrections.

Due to a technical error, longer text fields are cut off in CPR and NIIS in Access. This problem has been fixed by re-exporting from Rbase. Further inconsistencies between variable names in the written forms and the corresponding database fields were minor in both CPR and NIIS. Existing questions with empty database fields were automatically detected in the recoding process and were eliminated. Database fields with values, but without associated questions, in some cases could be deduced by checking other databases with identical variable names. Since the IFRI database is reviewed by three database experts at the Workshop for Political Theory and Policy Analysis in Bloomington and at the University of Michigan, no further checks have been performed for it.

Finally, the data was converted into Postgresql. Further processing and calculations were then carried out in R (http://www.r-project.org). The number of variables used in analysis is 401 for CPR, 404 for NIIS, and 705 for IFRI. After these steps, the three databases are available in a consistent and uniform internal format for further processing. All possible sources of error are checked and corrected. All further steps are now performed on these raw data, first the selection of the relevant variables.

3.5.2 Selection of data

Not all data available in the databases is relevant for the research questions discussed in the introduction. Excluding variables is usually trivial, since most variables in doubt are ID fields or empty fields. However, there are also variables that are discarded for reasons of content. One example is "What is the elevation of the intake?"

A fundamental question pertaining to selection of variables is whether a success factor can be better expressed by one or few key variables, or whether a more differentiated picture can be achieved by many suitable variables. This is ultimately expressed in the model quality achieved. In contrast to many other works (Gibson, Williams, and Ostrom 2005; Andersson and Agrawal 2011; van Laerhoven 2010), which characterize a concept such as social capital with only one or a few variables and usually measure success through an assessment of the forest condition by residents, this analysis chooses a much more differentiated way to operationalize individual success factors.

First, it is assumed that the combination of many variables yields a more accurate picture. Thus, each of the selected variables is assigned to one of the identified twenty-four success factors (see Section 3.5.3). For an overview of all success factors, see Section 5.1.3. Variables are assigned to indicators, that is, sub-factors within a success factor.

Second, for ecological success, the quantitatively collected data on the actual state of resources (see also Section 4.2.6) is used, if available, not the subjective assessment of local individuals. This results in a much more objective and accurate picture of sustainable use, although there are similarities between these methods. However, these could only be determined after a complex quantitative calculation (Salk, Frey, and Rusch 2014).

Since the first model runs have shown that IFRI models perform better if only a few key variables are used, but CPR and NIIS models perform better with the differentiated approach (more variables), all models are calculated for all data set variants (see Section 3.5.5). All models are presented and discussed in the results section.

The percentage of variables not used is 32% in the CPR (196 of 593), 27% in the NIIS (151 of 554), and 20% in IFRI (173 of 878). The comparable order of magnitude indicates that about one third in all three databases is database-specific information like IDs or empty fields. Thus, the proportion of variables used is 80% (705) in IFRI, 68% (401) in the CPR, and 73% (404) in the NIIS. After selection of relevant variables, the next step is recoding them into unified values that can be read by machine learning algorithms.

3.5.3 Recoding of variables

Before the data can be used, it must be recoded in order to make it machine-readable. This is the most time-consuming step. The aim is to bring all variables—whether text, date, number, or Likert scale—into a uniform, computer-readable numerical format. In particular, the neural networks used have specific requirements: the values must be between -1 and 1. Recoding all data to within the same interval also avoids biasing the neural networks toward large factors.

Special importance is given to the fact that the recoding is comparable across databases. This is achieved by always encoding neutral values with 0, positive extreme values with 1 and negative extreme values with -1, regardless of the actual distribution of values within a variable even if it is skewed. The relevant characteristic values for all factors are discussed in the results section and the Appendix, Tables 7.5, 7.6, and 7.7. Since many questions overlap across databases, such variables are recoded identically. Thus, 55% of the variables between CPR and NIIS are recoded exactly the same. The proportion of CPR/NIIS to IFRI is approximately 20% each.

Recoding involves an a priori decision on what is considered "good" or "bad" for the ecological success. For example, for social capital, 1 is equated with "very high social capital" and -1 with "very low social capital." This seems uncontroversial for factors 2, 3, 4, 5, 6, 9, 12, 14, 15, 16, 17, 18, 19, 20, 21, 22, 23, 24, since "more fairness" or "better information" should, in general, have a positive influence on ecological success. However, for 1 (resource size), 7 (number of actors), 8 (group composition), 10 (dependency on resource), 11 (dependency on group), and 13 (participation), a theoretically informed preliminary decision must be made. Based on the literature discussed in Section 5.1, a small resource size, a small number of actors, a homogeneous group composition, a high dependency on resources and the group as well as high participation possibilities are thus characterized as positive (1) and the respective opposite as negative (−1). Note that even if a wrong preliminary decision is made here, this would still be without consequence for the calculation of models, since only the sign would be reversed, but not the pattern recognized by the machine learning algorithms.

A preparatory step towards recoding is to separate groups of values in a database field (e.g., "01000" or "RYNR6") for meaningful further processing.

3.5.3.1 Recoding of variables—step 1: Aggregation

In some cases, it makes sense to group variables together. The first reason is content-related—if variables are closely content-related. Two examples taken from the CPR, NIIS, and IFRI database may illustrate this first kind of aggregation: the three variables: "physical sanctions by other appropriators," "social sanctions by other appropriators," and "sanctions by officials" are dimensions of criminal law that are often likely to be used together. They are combined into a dummy variable "Penalties." This makes the aggregated variable more meaningful and there are fewer missing values. The somewhat finer division into three different penalties is, however, lost.

Another example is variables that describe the state of the resource at the beginning: the appropriation resource, the production resource, and the distribution resource. They are also combined to a single variable which describes the overall condition of the ecological conditions well.

The second reason for aggregating variables is technical—if a variable is not immediately usable. In such cases, combining similar variables may be the most precise approximation. An example of this is the question of the duration of the control system. It can be calculated quite precisely using a combination of variables indicating, respectively, the start and end of the rule system, of the organizational structure and the dates for the location of the system. In addition, these combinations make it possible to compensate for missing data.

In order to work as closely as possible with the collected data, combinations are limited to the minimum. For the CPR data set, 290 variables are used for the final analysis, that is, the 401 variables result in 290 individual and combined variables. In the NIIS database, the 404 variables used in total result in 208 variables, some of them combined. For IFRI, 675 of the 705 variables used are partly combined. Table 3.1 gives an overview.

The relatively high number of combined variables, although limited to a minimum, is explained by the fact that a few combinations consist of very many variables. An example for CPR is success factor 1 (resource size), which fuses all 17 variables that make statements about the size of the system into one. Further examples are the success factors group limits (F12) and information (F16), where 16 and 18 variables are combined, respectively. For the latter two factors, the combinations come about through a rule matrix of individual rules. They are then summarized as to whether and to what extent it is possible to speak of an overall rule set. Which rule exists in detail is irrelevant for the success factor (Ostrom 1990). Therefore, all possible access restrictions are combined into one variable.

Table 3.1 Percentages of variables used in the three databases

Database	Variables in total	Used	Combined	% used
CPR	593	401	290	68%
NIIS	554	404	208	73%
IFRI	878	705	675	80%

Furthermore, the coding of the dependent variable (environmental performance) in NIIS is responsible for 26% (50) of all combined variables—again summarizing large matrices: water supply is measured according to season (spring, monsoon, winter) and simultaneously at the head and tail end of the irrigation system. In addition, a distinction is made depending on the product cultivated (rice, wheat, maize, vegetables, others, fallow land). All these individual variables are combined to form a meaningful key figure for water supply. Table 3.2 gives an overview for NIIS as an example.

Table 3.2 Combination of variables in NIIS

Success factor	Single variables	Combined variables	Used for combinations	Not used
F1—Resource size	17	2	17	3
F2—Resource boundaries	5	2	4	3
F3—Accessibility	3	0	0	0
F4—Ecological success at the beginning of observation	3	1	2	5
F5—Manageability	9	1	2	2
F6—Regeneration of RU	4	0	0	0
F7 –Number of actors	4	1	4	2
F8—Group composition	10	1	9	0
F9—Social capital	8	0	0	1
F10—Dependency on resource	13	2	3	8
F11—Dependency on group	4	0	0	1
F12—Group boundaries	22	2	16	0
F13—Participation of users	27	6	11	9
F14—Legal certainty and legitimacy	22	4	10	5

Continued

Table 3.2 *Continued*

Success factor	Single variables	Combined variables	Used for combinations	Not used
F15—Administration	12	0	0	0
F16—Information	25	3	18	2
F17—Characteristics of rules	29	2	15	6
F18—Fairness	12	1	2	0
F19—Control	11	1	4	0
F20—Compliance	19	4	12	0
F21—Conflict management	4	0	0	0
F22—Exclusion	3	0	0	0
F23—Relations	19	2	9	9
F24—Capabilities to adapt to change	9	1	5	1
25—Ecological Success	58	10	50	18

3.5.3.2 Recoding of variables—step 2: Text variables

Of the variables used, 30 in NIIS, 29 in CPR, and 313 in IFRI are text variables. Mostly, these are descriptions and comments. Recoding is done as follows. First, based on the information in the respective text variable, it is assigned to the appropriate success factor. Second, the textual information is transformed into a numerical value. Due to the subjective interpretation of text, only three tendencies are distinguished—positive, negative, and neutral. For example, for the success factor 18, fairness, there are three numerical values: A value of 1 is assigned for all assessments and formulations that describe fair conditions; a value of -1 for descriptions of unfair conditions and a value of 0 for all neutral descriptions. Here are two examples: "Tail enders are not getting adequate supply, even during monsoon." This is recoded with -1. The text "Fair water distribution and cropped area over the years has increased" is recoded with a 1. For IFRI, the text was partly interpreted by different people. However, the agreement as to whether a text was positive, negative, or neutral was almost 100%, so that the validity of the evaluation of only one person at CPR and NIIS should also be given.

For some success factors, there are no or not enough text variables. They therefore consist only of numeric variables. For NIIS, 32% of factors are without text (1, 5, 7, 11, 15, 19, 22, 24), in CPR this pertains to 44% of factors (1, 3, 4, 5, 6, 7, 16, 19, 20, 22, 24), and in IFRI to 16% (1, 2, 3, 7). For all other success factors with usable text variables, a value is calculated from all these variables. Since each success factor consists of several sub-concepts called indicators with several variables assigned to each indicator, this final value from text variables is added as an additional indicator to the numerical indicators of each factor.

3.5.3.3 Recoding of variables—step 3: Multiple use of variables

A few variables are also assigned to more than one factor if the answer options provide information for several success factors. In detail this means for CPR, 16 variables are used twice, 4 three times, and 9 four times. For NIIS 17 variables are used twice, 3 three times, and 1 six times. For IFRI 178 variables are used twice, 23 three times, 3 four times, and 1 five times. The much higher use of double and triple variables in IFRI is mainly explained by the much higher number of text variables. This information and the many very diverse answer options can be used several times. For example, the variable that is used five times has twelve responses.

However, it is crucial that the same information is not used both for one of the independent variables (twenty-four factors) and simultaneously for the dependent variable (success). If this were the case, statistical independence would no longer be given. After recoding, this was checked in a further step and, for all rare double uses, only the more suitable one was selected, the other use deleted.

3.5.3.4 Recoding of variables—step 4: Imputation

Dealing with missing values is a difficult problem. There is no best method, but many possibilities to compensate for the lack of data. The possibilities range from the exclusion of all cases in which values are missing to the replacement with the mean value of all cases or the clustering of similar cases and the calculation of an average value per cluster using a regression analysis (Sarle 1997).

The separation into training and test data sets for all three methods opens up another method. The mean error, that is, the quality of the model, is a measure of how strongly the predictions deviate from the correct values. These predictions are influenced by a different treatment of missing values. The result shows which replacement method is the best, because the existing data remains the same, while the replaced data changes in each case.

Imputation is necessary, because both machine learning algorithms random forests and neural networks, only take full data sets as input. In most cases, the assignment of multiple variables to indicators, which are in turn aggregated into success factors, takes care of missing values, since existing values act as stand-in for missing ones.

For example, for the NIIS database, imputation is generally not necessary. On the factor level, only 38 out of 6,575 data points (25 factors x N = 263) were without value. These 38 values were replaced by the mean value of the factor. For the other two data sets, a somewhat more complex procedure was chosen: the missing values were calculated using regressions based on the existing values. This applies to 101 values (25 x N = 122) of 3,050 data points for the CPR and 597 values of 10,225 data points (25 x N = 409) for IFRI. Expressed as a percentage, this means that an estimate is used for 3% of CPR, 0.6% of NIIS, and 5.8% of IFRI data. Since this is a critical transformation of the data despite the low percentages, all models were calculated twice as a precautionary measure—with and without imputation.

Each success factor is encoded separately. All recoded and combined variables are stored with their assignment to the indicators. This allows a raw value to be calculated for each indicator. It is calculated by dividing the sum of the variables per indicator by their number. The raw value varies depending on how many values exist for the variables that make up the indicator. As an additional check, a scatter plot of each success factor is prepared.

After this recoding step, the data is available in a uniform, machine-readable format: variables that belong together in terms of content are combined, data matrices are resolved, irregular Likert scales are recoded consistently with the correct factor alignment, and text questions are converted into numerical values. The subsequent steps—weighting and splitting—are done on this cleaned data.

3.5.4 Weighting of variables and indicators

Just as some variables have no relationship at all to the success factors and are therefore excluded, relevant variables may not be equally relevant. Within each success factor, it therefore makes sense to weight the variables assigned to it according to their relevance. This is done by weighting them from 0% to 100%. Of course, every model is also calculated without any weighting. The subjective nature of this weighting is problematic. This problem is addressed in two ways. First, three raters carry out the weighting independently of each other. This minimizes subjective influences. There are several methods to measure the agreement of the raters, such as simple percentage agreement, Cohen's kappa or Krippendorff's alpha. Krippendorff's alpha (Krippendorff 2004) is considered the best choice, as it is suitable for several raters, different sample sizes, and different measurement levels of variables (Lombard, Snyder-Duch, and Bracken 2002). The result shows (Table 3.3, calculated according to Freelon 2013) that the agreement of these three raters is very high.

These very high values mean that the relevant variables can be separated very well from less relevant variables. The final weighting used is the arithmetic mean of the three values assigned by the raters.

Table 3.3 Krippendorff's alpha for the respective classifications three raters per decision (weighting of variables)

Database	CPR	NIIS	IFRI
Weighting for level of variables	0.84 (786 decisions)	0.78 (678 decisions)	0.84 (2757 decisions)

An additional check is carried out by training all three modeling methods on weighted and unweighted data as well as data weighted only at indicator level. It turns out that both weighting and imputation of missing values have an important influence on model performance. However, it is only through testing many different models in their combinations that this becomes visible at all. This justifies this procedure just described post hoc.

3.5.4.1 Weighting of the variables and indicators 2: Indicator weighting

First, weighting is done on the variable level. Take as an example factor 18, fairness. The variable with the highest weighting here is "In your estimation, are the rules-in-use perceived by members of this subgroup as fair?" It is included in the value of the factor fairness with 21%. The variable "Have the relatively worse off been deprived of their benefits from this resource or substantially harmed?" with 11.66%. One reason is that the first variable precisely defines the success factor fairness. In contrast, the second variable also asks about fairness, but affects only a subset. These and other reasons reflect the independent weighting by three evaluators. It is possible to calculate the models with any other weighting—for example, that of other experts—without any problems. Different weightings thus provide another sensitivity analysis, which is presented in Chapter 5, Tables 5.34–5.43 and discussed in the interpretation (Chapter 6).

3.5.4.2 Weighting of variables and indicators—step 2: Weighting of indicators

Just as not all variables are equally relevant for a success factor, so too are not all indicators equally relevant for a success factor. An example: Factor 19 (control/monitoring) consists of two indicators. Indicator 1 measures the monitoring of the biophysical resource (biophysical monitoring); indicator 2 measures the monitoring of other actors (social monitoring). Indicator 1 is considered more important than Indicator 2 and is therefore weighted more in the final calculation of the factor (63.3% and 36.7%). A fully weighted data set (see Section 3.5.4.1), consists of data weighted at the variable level as well as at the indicator level.

The relevance of each indicator is determined by three independent raters. The agreement of the raters is again very good (see Table 3.4, calculated according to Freelon 2013).

Again, weighting indicators might bias the data, since assessment is again subjective. Therefore, three raters independently weight the indicators. The high level of agreement between them (Krippendorff's Alpha α between 0.85 and 0.93) shows that the relevance of the individual indicators is classified as fairly equivalent. In addition, individual patterns of the data remain unchanged—data points are merely moved. This is relatively unproblematic for pattern recognition by machine learning algorithms.

In order to evaluate all different weightings, all are presented in the results section (Tables 5.34–5.43). For CPR, NIIS, and IFRI as a whole, the model grades of the data sets weighted at indicator level adjusted R^2 changes by -6%, +2%, and +-0% for regressions, by -1%, +-0, and +-0% for random forests and by -25%, +-0%, and +1% for neural networks (see Section 5.6). Except for one outlier in CPR, the changes are therefore very small. Compared to other data set variants such as imputation, these changes are therefore not very significant.

3.5.4.3 Weighting of variables and indicators—step 3: Selection of the Top 3 variables

If using many variables to express a factor, it may be problematic that despite weighting not all variables point in the same direction, that is, operationalize the concept adequately. Principal component or correlational analyses may alleviate the problem but with hundreds of variables it will always be the case that variables differ in expressing a concept.

Three scenarios are conceivable: First, the variables may be in a negative relationship to the factor, even though they belong to a concept in terms of content. An example: A large number of fully paid guards may indicate that monitoring is very good. However, the guardians may be inattentive or corrupt. Without this knowledge, it is theoretically appropriate to assign the variable to the factor monitoring.

Table 3.4 Krippendorff's alpha for the respective classifications three raters per decision (weighting of indicators)

Database	CPR	NIIS	IFRI
Weighting for level of indicators	0.88 (258 decisions)	0.85 (258 decisions)	0.93 (153 decisions)

Yet, with the additional information of corruption it may become apparent that the variable has no or may even have a negative effect on monitoring. However, this is not evident from the data. It is as well possible that many different variables per factor, even if they point in the same direction, could take away each other's signal strength due to their fluctuations and different levels of intensity. The result would be a noisy signal. Third, it is possible that individual variables are negatively correlated, even if they match content-wise.

For these reasons, a variant is calculated for each data set that only takes into account the three variables that are weighted the highest per factor. These three key variables should definitely have a positive correlation and a clear interpretation of the content. This variant is referred to as CPR/NIIS/IFRI Top 3 in the sections that follow. This measure addresses the problems just mentioned. It also adds another comparison.

First, the raw data for CPR, NIIS, and IFRI are adjusted to an identical column format. Checks are made for correct sequence and non-unique variable names. In accordance with this scheme, a transposed weighting matrix of the three raters is created from which each rating can also be extracted individually. Thus, two data sets are created, one with raw data, one with weights corresponding to each other.

A R-script then multiplies the raw data matrix with the weighting matrix and creates a data frame for each variant. For each record, the weighting steps are checked to exclude formula and processing errors.

After these two steps (imputation and weighting), each data set exists in six variants—imputed and non-imputed (only for multivariate linear regressions (MLR)), as well as fully weighted, weighted at indicator level, and unweighted. This is supplemented by some special variants, for example, IFRI with individual weights or NIIS without revisits.

3.5.5 Split of data sets in training and test sets

In each database and for each analysis method (see Section 4.1), the data sets are split into a training set of 80% and a test set of 20% (in machine learning terminology, this would be called a *validation set*). This division is a standard procedure in machine learning to estimate the forecasting quality of the models for unknown data. The algorithms (here: the random forests and neural networks) learn from the training data sets by approximating the target function from input (independent variables) to output (dependent variables) over many repetitions. In the training set they are provided with both input (success factors) and output (ecological success) in order to match them. In the test data set, however, the value of the dependent variable belonging to a set of known independent variables is unknown (but known to the analyst). With this new input the algorithms have to predict the unknown output. The average deviation from the actual value (known to the model builders) allows analysts to calculate predictive accuracy. This allows analysts to estimate model quality for completely unknown data sets in which the true values of the independent variable are not known.

Like other parameters, the composition of the training data set plays a major role for model quality. Therefore, sorting the data according to its distribution of one success factor is a first step (e.g., size). It is now conceivable, for example, that large systems are on average more unsuccessful due to their complexity. A simple split by size—using the 80% smallest systems for training and the 20% largest systems as tests—would lead to bad (here: too positive) predictions. Therefore, different splits are used. These have been mentioned or used in previous research to separate systems in analysis (e.g., Tang 1989): size (success factor 1), social capital (success factor 9), external relationships (success factor 23), and success (output). A random separation is used as a check. In addition, selection after sorting (every fifth case) ensures that the data is balanced in both training and test data set, that is, in the example both small and large systems are contained in the training and test data set.

Further splits are according to the type of administration (agency-managed irrigation system (AMIS) vs. farmer-managed systems (FMIS)), and after intervention of a system by the government or non-governmental organizations. The ratio of training to test data is always 80% to 20%. For this split, care is taken to ensure that all systems are in the training data set before state intervention and all systems are in the test data set after the intervention. An "X" in Table 3.5 means that this division has been made for this data set. Not all splits have been performed. For NIIS, for example, the division by country is pointless, since all cases come from Nepal.

Further splits are based on each factor per data set with the highest correlation with the dependent variable, ecological success. This is the factor Fairness (F18) for CPR, factor Participation (F13) for NIIS, and factor Ecological Success at the beginning (F4) for IFRI.

For NIIS, an additional data set was created which does not include revisits (n = 244 instead of n = 263) in order to avoid the problem of autocorrelation in the statistical analyses used. Further splits were also tested, such as a random split. None of these are listed in the results reported here, since their model performances were very weak.

Table 3.5 Split criteria for the four data sets

Name	Criterion: Guarantees that training set and test set are balanced in regard to . . .	CPR	NIIS	IFRI	IFRI Top 3
Relations	. . . the distribution by relations with other parties (factor 23)	X	X		X
Size	. . . the distribution by size (factor 1)	X	X	X	X
Ecological success	. . . the distribution according to ecological success (factor 25, dependent variable)	X	X	X	X
Social capital	. . . the distribution according to social capital (factor 9)	X	X	X	X
Revisit	. . . the distribution after repeated visits		X	X	
Ecological success at the beginning	. . . the distribution according to ecological success at the beginning (factor 4)			X	X
Fairness	. . . the distribution according to fairness (factor 18)	X			
RU-Type	. . . the distribution by resource type (fishing or irrigation)	X			
AMIS/FMIS	. . . the distribution by management type (agency or farmer managed)		X		
Participation	. . . the distribution according to participation possibilities (factor 13)		X		
Country	. . . the distribution according to the country of origin of the case studies			X	

3.5.6 Preparation of the analysis results

After splitting, recoding is complete. Data are available in different variants for statistical analysis. Table 3.6 shows schematically how a ready prepared data set looks at the success factor level.

For statistical analyses, data sets follow a specified naming convention. This is done on a file level via batch commands. This also ensures that they are copied to the corresponding folders. In the analyses themselves, a script programmed in Java executes the batch processing and splits the data sets into folders according to split, model parameters like repetition rate, learning algorithm, or weighting.

Table 3.6 Format of the recoded data for reading into neural networks, numbers rounded for display

Case	1 Size of resource	2 Boundaries of resource	3 Accessibility	4 Ecological success at the beginning	5 Manageability	6 Regeneration of RU	7 Number of actors	8 Group composition	9 Social capital	10 Dependency on resource		25 Ecological success
1	0.51	0.93	-0.54	-0.54	-0.06	0.10	0.90	0.73	0.29	...		0.30
2	0.87	0.93	-0.54	0.37	-0.06	0.10	0.82	0.50	0.35	...		0.69
3	0.87	0.93	-0.48	0.37	-0.06	0.28	0.82	0.60	-0.10	...		-0.12
4	-0.76	1.00	0.77	-1.00	0.06	0.06	-1.00	0.72	0.64	...		0.03
5	0.92	0.50	-0.54	1.00	0.58	0.52	0.70	0.39	0.05	...		-0.15
...	...											...
122	0.83	1.00	1.00	-0.50	0.89	0.00	0.93	0.36	0.74	...		0.31

With this workflow, it is possible to automate the execution of many networks (>70,000). However, the network architectures must be created manually. The number of neurons in the hidden layer varies with constant input (success factors) and output (ecological success).

The results are evaluated in three R-scripts, one for regressions, one for random forests, and one for neural networks. Here, the target data for ecological success is compared with forecast results; characteristic values of model quality are automatically calculated using the deviation from true values. For regressions, this is done via the *fitcorr* function, for the random forests via the *importance* function, which measures the increase in node purity (see Section 4.1.2), and for neural networks via the *mean absolute error* function. Sorting according to this criterion determines the best models.

3.5.7 Summary of methodology and data

Three large available data sets on SES meet the high demands for an analysis of ecological success and possess enough information on twenty-four independent success factors. All data sets originate from the Workshop for Political Theory and Policy Analysis in Bloomington and were largely developed by Nobel Prize winner Elinor Ostrom. Together, they present 794 cases from three sectors— fisheries, irrigation systems, and forest management—which provide the characteristics of groups of actors involved as well as an extremely detailed picture of biophysical conditions, and social rule systems. They are probably among the highest quality large-N-studies in the field of SES available (Poteete, Janssen, and Ostrom 2010).

This section (Section 3.5.7) described the individual working steps that were necessary in order to arrive at an assessment of the complex interactions between the various presumed influencing factors from these raw data. On the one hand, this means recoding the raw data and allocating individual variables to twenty-four potential success factors. All critical steps in this process are documented in Section 3.5.3. In addition, for each critical decision (such as imputation or not; weighting or not), the respective models are presented in the results section so as to be as objective and transparent as possible. These variants contribute in addition to robustness and sensitivity analyses, since it is always possible to see the model quality and the respective role of the influencing factors for a specific model—always in comparison to the raw data which is neither weighted nor imputed.

The various machine learning algorithms used also contribute to this methodological caution. Models with different methods—multivariate linear regressions, random forests, and neural networks (see Section 4.1)—are deliberately calculated for all data sets. This allows not only a better assessment of the strengths and weaknesses of each model, but also to track the robustness of the models using different methods. It will become apparent that the results vary considerably in some cases.

Another step is to develop a validated indicator system. It must be able to operationalize influencing factors as described in the literature consistently and across databases (see Section 4.2).

4

Methods

This part introduces the methods used. The three used statistical methods are multivariate linear regressions, random forests, and artificial neural networks. For these statistical analyses the open source statistics software R (https://www.r-project.org/), version 3.3.4 is used.

A further methodical step is the *operationalization* of success factors. This is done through the development of a system of indicators (see Section 4.2) which are theoretically derived in Section 5.1 and examined with regard to their influence on ecological success.

4.1 Introducing the three statistical methods used

This analysis uses three different analytical methods as a methodological precaution: multivariate linear regressions, random forests, and artificial neural networks. The main reason is to be able to estimate the robustness of results independent of method, since data complexity is very high. A second reason is that no model assumptions can be made before the analysis, and it is unknown which patterns are present, thus it is to be determined which machine learning algorithm performs best. Third, each of these methods has specific advantages and disadvantages. The use of several methods allows analysts to avoid the respective weaknesses and thus achieve reliable results. I will start with linear regressions, which are mainly used as a well-known baseline.

4.1.1 Multivariate linear regressions

Multivariate linear regressions (MLR) are perhaps the most frequently used analysis method when it comes to assessing simultaneous influence of different independent variables on a target variable. The general regression equation is

$$y = b_0 + b_1 * x_1 + b_2 * x_2 + \cdots + b_n * x_n + u$$

Sustainable Governance of Natural Resources. Ulrich Frey, Oxford University Press (2020). © Oxford University Press.
DOI: 10.1093/oso/9780197502211.001.0001.

where y is the dependent variable (here: ecological success) and x_1 to x_{24} are the potential success factors F1 to F24, and u is the error term.

When estimating the regression coefficient, the sum of squares of residuals is minimized using the least squares method. The model quality is given as adjusted R^2. The advantage of MLRs are their high robustness, even in case of violations of the model requirements, their applicability even in small numbers of cases and their high dissemination, which facilitates comparisons. On the other hand, MLRs tend to assume linear correlations between target variable and independent variables, even if there are none. Unfortunately, it is exactly this non-linearity that is assumed in the data sets used in this study. This does not mean that regressions are unsuitable at all, it only means that they cannot predict anything outside linear relationships (for further details on regressions, see Backhaus, Erichson, and Weiber 2013). For this reason, regressions are mainly used in this study as reference and baseline to put the model quality of the other two methods into perspective, since model performance is expected to be inferior due to being constrained to modeling the linear relationships only (for further details on regressions, see Backhaus et al. 2008).

4.1.2 Random forests

Decision trees are a well-known machine learning algorithm, often used for classifying large amounts of data. Decision trees are used when the model is unknown, the complexity is high, and a lot of data is available. All data sets used in this analysis have this structure. Like regressions, decision trees are used frequently, since decision rules can be understood as simple if-then rules, which allows a simple interpretation of even complex data sets (Mitchell 1997; Clarke, Fokoué, and Zhang 2009). These rules represent the most important characteristics of the data set under investigation (Alpaydin 2010; Myatt and Johnson 2009).

Like neural networks, decision trees belong to supervised learning models. Supervised means that the output (the dependent variable) is known.

A decision tree is a hierarchical data structure implementing the divide-and-conquer strategy. It is an efficient nonparametric method, which can be used for both classification and regression. (Alpaydin 2010, p. 185)

A decision tree uses decision rules to split a data set from a root node into additional nodes. The last level, the leaves, represent the last classification step. The structure of decision trees is not predefined at the beginning, but is generated by the algorithm itself (Alpaydin 2010). Figure 4.1 shows an example of a simple tree.

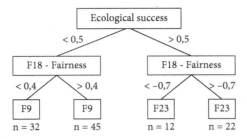

Figure 4.1 Example of a decision tree

Decision trees have numerous advantages. First, they are robust against outliers and can also be used for binary or ordinal variables. They are also suitable for non-linear relationships, which are suspected in the data. Second, in comparison to other machine learning algorithms, they perform well across many different problems and data sets, although they often perform slightly worse than neural networks. Clarity may count as a third advantage—the importance of independent variables can be seen at a glance (higher up in the tree equals a higher importance), specific interdependencies of the variables can be traced easily (succession of variables split) and data set partitions are easy to grasp. In our fictitious example above, factor 18 (fairness) is best suited to divide the data set in terms of ecological success. All cases with a value <0.5 are sorted left, those with a value >0.5 right.

The bottom line, that is, the leaves (here, e.g., n = 32), shows how many cases are assigned to each leaf. Thus, the sequence of decision rules in our example is that there are 72 (32 + 45) cases with low success (left side, i.e., <0.5). Of these, one cluster of 32 has both low fairness (<0.5) and low social capital (F9<0.4).

Disadvantages of decision trees are slightly lower quality compared to other classification methods such as neural networks. Individual decision trees often create models that are too simple or too specific. However, this can be improved by using many trees (random forests or RF). Their majority decision is much closer to the real result than a single tree (Williams 2011). These forests then consist of about 100 to 5,000 unpruned trees, each of which is overfitted. However, this is rebalanced by the majority decision of the ensemble of decision trees. In random forests, the majority decision applies; that is, a case is sorted into the category in which most trees place it. If the goal is not classification but a precise estimation of separation values (regression forests, as it is here), then the mean value of the trees is used (Williams 2011).

Like other machine learning algorithms, specific parameters of random forests can be tuned in order to improve their performance. Two parameters are especially important—the number of trees and the number of parameters used for each division into branches of the tree (*mtry*). Hence, the proposed standard value in R (*randomForest*) is not adopted. Instead, the best value for each split (of each data set, see Section 3.5.5) is first calculated using a function (*tuneRF*). However, manual checks revealed that this tuning function did not result in optimal model performance. Model performance was optimal when then *mtry*-parameter was either 4, 8, or 16. Therefore, as a precautionary measure, each permutation of the parameter (4, 8, 16) with every number of trees (500, 1,000, 1,500) was calculated for each data split. The model with the highest explanatory power was adopted. The models and their parameters can be found in Tables 5.2 (CPR), 5.10 (NIIS), 5.18 (IFRI), and 5.26 (combined model).

Apart from the advantages and disadvantages mentioned, random forests do have one quality which is very important to this study. They are very robust against noise, that is, limiting the effect of useless information on model quality. In this study, we assumed that some of our twenty-four success factors would not have a verifiable significance for success. In fact, they might prove irrelevant. Therefore, we needed robust, rather well-performing models that are able to cope with irrelevant factors and allow to estimate the importance of each success factor (Breiman 2001). Random forests do exactly this which is mainly due to the large number of individual trees and to the optimized algorithms on which the software packages—implementing the work of Breiman (2001)—are based.

In comparison, neural networks are not able to directly quantify the relevance of independent variables. There, they have to be determined indirectly (see Section 4.1.3.4). In random forests, the criterion for the significance or relevance of a factor is the percentage increase of the mean squared error (MSE) at this variable's random change. If a variable is not important for predictive accuracy, the MSE and node purity (probability of misclassification) should change only little with random changes (Williams 2011). If the variable is important, the error increases significantly. The importance of each factor is reported for each random forest model (see Tables 5.3, 5.11, 5.19, and 5.27 in Chapter 5).

4.1.3 Artificial neural networks

For the problem at hand, there is agreement that potential success factors form a complex network of relationships. Similarly, it is considered certain that the contribution of some factors is not linear (e.g., size of the group; in this case, a medium size is considered optimal). The mutual amplifications and attenuations of the factors among each other are also non-linear. However, it is completely unclear in which interactions the success factors are related to each other.

These findings contrast with a lack of methods to address these problems analytically. Qualitative analysis is usually limited to some few cases. Although these can be analyzed in detail, they do not allow generalization and are restricted to some configurational patterns.

In quantitative analyses, linear regressions are common. However, most case studies only consider a few variables to be included in the model (e.g., Chhatre and Agrawal 2008; Waylen et al. 2010). This has several serious disadvantages, however: On the one hand, it is known that significance, direction, and strength of the effect of a variable can change dramatically if other variables are included in or removed from the model. Thus, the significance of such models is limited to exactly this combination of variables—again, a generalization is not possible. On the other hand, these variables are supposed to be in a linear relationship, which de facto is not the case. Finally, the number of variables considered rarely exceeds ten, but it is certainly higher. This can be deduced from the large number of independent variables used, which can predict different measures of success in individual systems with partially satisfactory predictive accuracy. Furthermore, multivariate linear regressions are only conditionally suitable for more complex models.

Neural networks allow to solve these problems, which is exactly the motivation for choosing them, since the primary goal of this investigation is to model the postulated complex interdependencies (Agrawal 2002; Tucker 2010; Lam and Ostrom 2010) between potential success factors and the actual success or failure of SES. Artificial neural networks seem to be the best tool for quantifying possible non-linear connections in a model with unknown model structure, since they are known to deliver peak model performance for pattern recognition in large data sets.

Neural networks are well-known, non-parametric statistical procedures. Originally developed for the investigation of biological nervous systems, they possess an analogous structure and represent an information-processing structure capable of learning from external influences. They allow complex modeling even of non-linear systems. Neural networks are often superior if the relationship between variables is either unknown or very complex (for a general statistical introduction see Backhaus, Erichson, and Weiber 2013, for a technical introduction Sarle 1997). Therefore, there is no need to postulate a correlation between dependent and independent variables. An advantage is therefore that no causal hypotheses regarding the relations of the variable must be taken into account (Backhaus, Erichson, and Weiber 2013).

Neural networks can be used to model complex relationships between input values and output values, that is, to identify patterns in data sets. In many other areas, neural networks have already been successfully used as pattern recognition methods. Examples include cancer cells typology (Khan et al. 2001), facial recognition (Rowley, Baluja, and Kanade 1998), analysis of microarray data in genetics (O'Neill and Song 2003), climate modeling (Knutti et al. 2003), and other applications (Widrow, Rumelhart, and Lehr 1994). There are also some examples for natural resource management (Peng and Wen 1999).

Neural networks have two disadvantages. On the one hand, they require large amounts of data. On the other hand, they only provide an implicit, not an explicit model. They ultimately remain a "black box," in the sense that the extraction of a closed function is not possible according to the current state of research. Thus, the respective learned connection between input and output signals cannot be interpreted. Nevertheless, the model represented by the networks is fully predictable: parameter changes in input (independent variables) result in changes in output (ecological performance). This makes them suitable for problems where the result (model quality) rather than the nature of the relationships is of interest. Yet, recently, there have been advances in extracting this information from neural networks (Thrush, Coco, and Hewitt 2008; Yeh and Cheng 2010). This analysis implements these theoretical methods in order to estimate the importance of each success factor for the sustainability of SES.

4.1.3.1 Construction of artificial neural networks

ANN are simple replicas of the nervous systems of living beings. They consist of neurons that can exchange information with each other via connections. Each neuron receives any number of connections as input and outputs a value as output. This value can be passed on to any number of neurons. There are many different types (architectures) of neural networks (cf. for example LeCun, Bengio, and Hinton 2015; Sarle 1997; Mehrotra, Mohan, and Ranka 1997).

For neural networks, the existing data quantity and structure is relatively simple. Therefore, as architecture, feed forward networks are chosen. In these networks there are no feedback loops from outputs to inputs. The latter consist of a layer of input neurons and read in the data for the twenty-four success factors of SES. They are connected to a hidden layer of neurons, which in turn are linked to an output neuron that represents the network's predictions of ecological success. Figure 4.2 shows this design.

In addition, all networks used are fully connected. This means that each neuron in the hidden layer (here H1–H5) receives its input from all neurons of the input layer (here the twenty-four input neurons RS1–EE3) and passes the output on to all neurons in the output layer (here ES; see Reed and Marks 1999). The output of a neuron is calculated from the values of the incoming connections and the activation function in the neuron itself (see Figure 4.3). The weightings of the input influence the activation of the neuron in strength and direction. The activation function establishes the connection between input and the activity of neurons. A negative sign means inhibition, a positive sign means amplification.

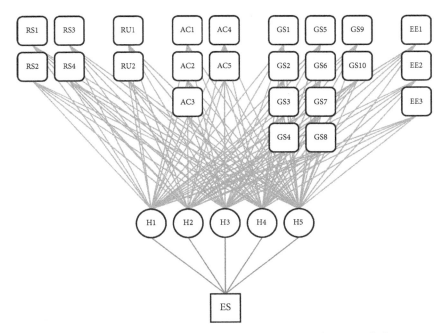

Figure 4.2 Neural network with twenty-four inputs neurons (RS1–EE3), five hidden (H1–H5 = hidden), and one output neuron (ES = ecological success)

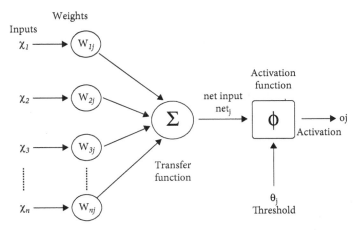

Figure 4.3 Scheme of weighting, activation, and transmission in neural networks

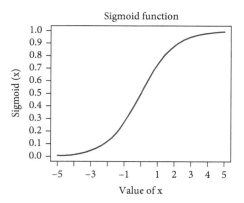

Figure 4.4 Sigmoid activation function in neural networks

There are various activation functions. For this analysis a sigmoid (logistic) function is used (see Figure 4.4).

This sigmoid function is non-linear, monotonous, and continuous (Reed and Marks 1999). It can be influenced by the threshold value. At very large and very small values it is saturated, at the zero point it behaves almost linearly, that is, for large negative values it becomes 0, for large positive input values it becomes 1. As equation it is expressed as follows (Reed and Marks 1999, p. 315):

$$y(u) = \frac{1}{1 + e^{-\lambda u}}$$

In addition, the activation function *tangenshyperbolicus* is used (Reed and Marks 1999, p. 315):

$$y(u) = \tanh(\lambda u) = \frac{e^{\lambda u} - e^{-\lambda u}}{e^{\lambda u} + e^{-\lambda u}} \tag{1}$$

As with the sigmoid function, the sum of all incoming inputs is used as basis for activation, but the tangent hyperbolicus function is steeper and its result is between −1 and 1, while the logistic function returns a result between 0 and 1. With this architecture, a neural network is—mathematically speaking—a non-linear, statistical method for approximating functions of any complexity (Reed and Marks 1999; Backhaus, Erichson, and Weiber 2013).

4.1.3.2 Learning and generalization

During training, the networks adjust the weightings of the connections between the neurons step-by-step to find the best fit between the actual values for ecological success (as specified in the training set for neural networks) and the values currently predicted in the training phase. Figure 4.5 shows the target function (solid line), which should be approximated as accurately as possible. The characteristics of the solid line represent the actual values in the training set of a database for the ecological success of all case studies. Starting from a randomly selected initial function (dotted line), the neural network now tries to adapt this function so that it corresponds to the actual values as exactly as possible. This is done by a stepwise adjustment which varies depending on the learning algorithm used and the number of repetitions.

As soon as this training is completed after a given number of repetitions (called epochs), the trained nets are validated by predicting the unknown success of the test set. The number of epochs is also varied, since this is also important for predictive success.

The learning process proceeds as follows: First, the initial weights are initialized with random values. Then, the value of the output neuron is calculated for each case. The average error between predicted value and actual value over all inputs is calculated and minimized by changing the weights. Since the error is to be minimized, the required function searches in the opposite direction of the steepest slope of the current values, and it is calculated with the help of the gradient descent method. The learning rate defines the step size with respect to the change of weights.

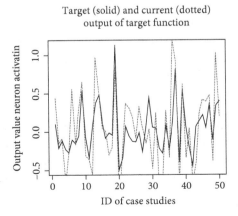

Figure 4.5 Training of a neural network—step-by-step approach to the target function

4.1.3.3 Design

There is no procedure for selecting the optimal neural network layout for a specific problem. The most suitable design can only be found by trial and error. This results in an extremely large number of permutations that have to be tested by brute force. Many parameters can be changed, which is called hyper-parameter tuning:

- the number of hidden layer neurons,
- the training method,
- the learning rate,
- the number of epochs in training, and
- the split into training and test sets.

The number of input and output neurons, however, remains constant. The number of input neurons (called *feature selection* in machine learning), the success factors, has been determined by a literature synthesis (see Section 5.1.5). This means that some potential factors may prove to be irrelevant. This can be taken into account by the neural networks.

The output neuron remains constant because only one dimension is chosen for ecological success. In theory, success could be split into an ecological, economic, and social dimension, which would result in three output neurons. However, there is not enough data available for the economic dimension of success. In addition, the social dimension is regarded as a prerequisite for ecological success, that is, as an input. There is also a methodological reason for choosing this architecture. The performance of neural networks typically deteriorates dramatically in the case of multiple output neurons (Reed and Marks 1999). The following paragraphs discuss which parameters are varied and why.

Number of neurons of the hidden layer
The number of neurons in the hidden layer determines the network's ability to grasp the required connections (function). In general, the larger the number, the better the performance. However, the number of neurons also increases training time and consumes computer resources. More problematic than this, however, is that with increasing computing power of the network (higher number of neurons), the generalization capability decreases, since the network then acquires the capacity to "learn by heart," that is, to memorize each individual pattern. The goal is to find the optimal trade-off: as few hidden neurons as possible, that is, a network that is just able to solve the task, but is forced by the number of neurons to generalize.

Training method
The training algorithms used are gradient descent methods. These are widely used for optimization problems. Starting from a random starting point, which is

achieved in neural networks by randomly weighting the connections at the beginning of the training, the aim is to change direction toward the smallest error. The gradient points to the highest error E. A change in the weights therefore takes place driven by the negative gradient. The gradient indicates the direction of the steepest descent from the respective previous value. This minimizes the error (Reed and Marks 1999, p. 57):

$$\Delta w_{ij} = -\eta \frac{\vartheta E}{\vartheta w_{ij}}$$

Formula 1: Weight change for gradient descent
Here w_{ij} is the weight change, η (Eta) the learning rate, and ϑE (Theta) / w_{ij} the derivative of the error rate regarding weights. If an input pattern is created and propagated forward through the net, the output of the net is compared with the desired output. The error results from the difference between the input vectors and their representation on the output vectors, that is, between target and actual. The quality of the agreement is measured by the error function:

$$E = \frac{1}{2} \Sigma_p \Sigma_i (d_{pi} - y_{pi})^2$$

Formula 2: Error function
P is the pattern in the training set, i the output neurons, and d_{pi} and y_{pi} are the target and the actual output for the i-th output neuron of the p-th pattern (Reed and Marks 1999, p. 52). The difference between the two values d_{pi} and y_{pi} is the error. This error is now propagated back to the input layer via the output layer. The weightings of the new neuron weights are changed depending on their influence on the error. This guarantees an approximation to the desired output when the input is created again.

The name of the algorithm results from feeding back the error to the network—error backpropagation. All three algorithms used (resilient propagation, backpropagation with momentum, and standard backpropagation) belong to this class of backpropagation. In addition, all of them are so-called supervised algorithms. This means that the networks try to achieve a correct output, which is given. The difference between the nominal and actual value of the given output vector (the error) is returned by the net (error backpropagation). The *momentum* of the second algorithm takes into account the last change of the weights and thus allows a faster approach to the target function in theory. In the available data sets, however, this algorithm achieved consistently poorer results than without momentum.

Standard backpropagation
This is by far the most commonly used method to train neural networks. The difference between the output of the current function of the network for a value and the actual target is minimized. Resilient propagation (RPROP) is a variant of backpropagation. This algorithm only takes into account the sign, but not the amount of the change in the error in the adjustment of the weights. The reason for this is the assumed non-linear correlation of the error surface, where the amount of change is not an indication for optimal improvement (Reed and Marks 1999). This algorithm consistently achieves the best results with the social-ecological data sets used.

Backpropagation with momentum
Here a momentum is added to the weight change towards the smallest error. Formula 3 is supplemented with a term that takes the last weight change into account:

$$\Delta w_{ij}(t) = -\eta \frac{\vartheta E}{\vartheta w_{ij}}(t) + \alpha \Delta w(t-1)$$

Formula 3: Backpropagation with momentum
If both terms point in the same direction, they strengthen and accelerate learning, if they point in different directions, the descent is slowed down to a minimum (Reed and Marks 1999, p. 71).

Learning rate
This parameter (see Formula 3) can only be changed for backpropagation with momentum and standard backpropagation algorithms. It varies the speed with which weight changes are made. The following applies: the lower the learning rate, the slower the network learns. If, on the other hand, the learning rate is too high, the probability that the weights and the function will diverge increases, so that there is no learning effect. Since the RPROP algorithm consistently achieves the best results with no adjustable learning rate, this variation is not very important. Values from 0.05 to 0.3 are used. These limits are within the recommended learning rates (Reed and Marks 1999).

Number of epochs in training
The number of epochs neural networks are trained proved to be a decisive parameter for the data sets used. Epochs range between 50 and 2,500. The following applies: the fewer repetitions, the higher the generalization ability, since the existing patterns are learned "by heart" with more repetitions. Since the best results were achieved with few repetitions, the number of repetitions in later runs was limited to a maximum of 350.

Splitting the training and test sets

To test predictive accuracy, that is, model quality, it is common practice for all classification algorithms to separate the data set into a training and a test part. Model adjustment takes place on the training set (80% of the data), whereas predictions of the model are tested on the unknown data of the test data set (20%). This allows analysts to quantify exactly the deviation of the trained model from the actual values. The deviation between predicted and observed data then gives the actual error rate and model quality (as adjusted R^2 or pseudo-R^2). These divisions are also carried out for multivariate linear regressions and random forests. The type of split is determined by considerations of content, such as success, size, or other factors. This is described in Section 3.5.5.

Technical implementation

With the help of a control program written in Java, the software MemBrain (MemBrain V03.08.01.00, http://www.membrain-nn.de/) trained various neural networks in different configurations on the learning data and determined and saved their predictions on the test data. The network architecture, learning algorithm, and number of training sessions were systematically varied (see Table 4.1) and saved alongside the results.

Table 4.1 provides an overview over the various components of the training process as well as the best parameters. Not all configurations were applied to all data sets, as some components quickly proved unsuitable (e.g., the selection of other activation functions other than the logistic one). All of these parameters were systematically varied. This resulted in around fifty network architectures and more than 55,000 configurations tested. Most testing was done on the CPR data set (43,065), with 2,089 on NIIS, 6,264 on IFRI, and 2,436 on the overall model.

Table 4.1 Training parameters of neural networks used

Component	Learning algorithms	Number of neurons in the hidden layer	Activation function	Number of epochs
Variations	(1) Resilient Propagation (RPROP) (2) Standard Backpropagation (3) Backpropagation with Momentum	1 to 28	(1) sigmoid / logistic (2) tanhyp	50–2,500
Best results	(1) RPROP	Medium number (8–15)	(1) sigmoid / logistic	Few epochs (50–150)

Since all runs were repeated with non-imputed data and different weights, another 20,000 model configurations were added to these base runs.

Depending on the scope of the configurations and data sets to be tested, this analysis took between six and forty-eight hours on a standard office PC.

To obtain an approximate measure of the robustness of the predictive quality of a certain network configuration, each configuration was trained three times and the corresponding three predictive results were saved. For each configuration of learning algorithm, number of epochs, etc., the training is therefore performed a total of three times. This allows analysts to check whether the training result remains constant (robustness, see Section 5.6). In this way, very good results due to a lucky strike can be ruled out.

There are four parameters in particular that mainly influence the quality of the networks in the data sets used here: (a) number of hidden neurons, (b) learning algorithm, (c) split into training and test set, and (d) number of epochs. The number of epochs is special, because very good training results may result in bad predictions in the test data set (overfitting). Overfitting describes the tendency of statistical models to describe noise or errors instead of real correlations and to react too strongly to small changes. The mathematical definition for this is:

Given a hypothesis space H, a hypothesis h element H is said to **overfit** [bold in the original] the training data if there exists some alternative hypothesis $h' \in H$, such that h has smaller error than h' over the training examples, but h' has a smaller error rate than h over the entire distribution of instances. (Mitchell 1997, p. 67)

Evaluation of the results

In a third step, using an R-script, the predicted output was compared with the actual values of the real cases in the test data set. Essentially, three indicators were calculated for the predictive quality: the mean squared error (MSE), the mean absolute error (MAE), and the determination coefficient (adjusted R^2). The latter was calculated according to its general definition with real data y_i, and forecast f_i:

$$R^2 = 1 - VAR_{residual} / VAR_{total} = \frac{\Sigma_i (y_i - f_i)}{\Sigma_i (y_i - \bar{y})^2}$$

Formula 4: Adjusted R^2

Finally, the arithmetic mean of the determination coefficients of the three predictions of each configuration was calculated as a measure of the robustness of a network configuration. Only these robust results are used in the discussion (Chapter 6).

4.1.3.4 Extraction of the relevance of factors

Although the extraction of the exact model from a trained neural network is still not completely possible, several methods have been developed in recent years that allow a more detailed analysis of the implicit neural network models. For these analyses, a sensitivity analysis method based on a comparison of relative weights between neurons is implemented (see, e.g., Gevrey, Dimopoulos, and Lek 2003). This method has proven itself in a rigorous comparison of different methods for sensitivity analysis in multilayer perceptrons (which corresponds to a circular-free graph whose edges are directed) (Olden, Joy, and Death 2004) and can be implemented relatively easily for the artificial neural networks used here. Following the outlined algorithm (Gevrey, Dimopoulos, and Lek 2003) an analysis routine in R was programmed, which calculates the relative influences of each influencing factors (input neurons) on the dependent variable (the output) from each neural network containing the weights of the trained network connections.

However, extracting the relevance for each factor using this method has its limits. This is due to a well-known problem of the different sensitivity analysis methods in multi-layer perceptrons: due to the initial stochasticity of the distribution of weights in the networks only relatively high influences (> approx. 15%) of individual inputs can be correctly detected. The results of individual sensitivity analyses are therefore only of limited significance for the problem at hand—it is evident that none of the influencing factors alone seem to have such a high influence. Reflection on the nature of SES already suggests this.

Before the described machine learning algorithms can be applied, it is necessary to operationalize both success factors and outcome, ecological success. This is done through the development of a system of indicators in the next section (Section 4.2).

4.2 Operationalizing the success factors via a new indicator system

4.2.1 Why do we need a new indicator system?

The success factors discussed (Section 2.5) are abstract concepts. They are more or less well defined in the literature (Frey 2017a). In the present case, data are available in the form of measurements or variables obtained by surveys and the task is to assign them to the success factors. *The concepts must therefore be operationalized.* The design principles and part of the SES framework have been operationalized by some studies, although differently (Leslie et al. 2015; Blythe et al. 2017; for a review see Partelow 2018).

The most common solution is probably the use of an indicator system (for local systems see, e.g., Boyd and Charles 2006; for the national level, see OECD 1994). An indicator system takes on the role of a mediator between measurable data and abstract, sometimes non-measurable concepts. It also ensures that different research groups collect the same data, since it is precisely defined which data must be collected or measured for which concept.

If concepts or indicators are used differently between research groups, comparability suffers. Thus, for example, the conceptualization used here includes the presence of a leader or a group (leadership) in the concept of social capital. Some authors, however, discuss these two concepts separately. Different conceptualizations or definitions become problematic as soon as measured data is assigned to these concepts, since a different understanding leads to different results. If different data sets are to be checked for one criterion, for example, forest management on the ecological condition of the forest, then this is only possible if the same indicators have been used.

In some cases, this will still not be possible. If, for example, the condition of a forest is measured by the indicator of its biomass, a simple comparison has to fail, because a healthy tropical forest and healthy tundra vegetation differ massively in this respect. This problem can be avoided by using reference points, which is exactly what has been done with the IFRI data (Salk, Frey, and Rusch 2014).

In order not to expose oneself to be criticized of subjectivity when assigning indicators to variables and concepts, questions such as "Which variable belongs to which success factor?" or "Do some variables express one concept better than others? Should they therefore be given a higher weighting?" have to be addressed through reproducible, systematic, and reliable procedures. These have been described in Sections 3.5.2 and 3.5.3. Other parts include transparent data handling, interrater reliability (see Section 3.5.4) and a full description of the indicator system used.

4.2.2 How to develop and validate an indicator system

There are many indicator systems (e.g., Böhringer and Jochem 2007; OECD 1994; overview in Lammerts van Bueren and Blom 1997). Nevertheless, to my knowledge there is no generally accepted or standardized system for SES for the scale of user groups (communities) using CPR. This may be due to the fact that most indicator systems operate on a national or even global level. The regional models available are still too imprecise for the purposes of this work. In addition, most indicator systems are limited to biophysical and demographic indicators. For SES, the entire social indicators, including rule systems, are thus missing.

For these reasons a new indicator system was created. Three quality criteria in particular distinguish indicators in general—validity, reliability, and availability (Boyd and Charles 2006; Tucker et al. 2008). A *valid* indicator actually measures what it claims to measure. If it is *reliable*, it measures this objectively, that is, comparably. Finally, the indicators must be ascertainable or measurable at acceptable costs, that is, *available*. This is always a question of costs and benefits and ultimately a compromise. In principle, perfect indicators are possible. However, costs would then be immense.

The indicator system developed (Section 4.2.3) uses the classification of the SES framework as first level. This results in five categories (labeled A, B, C, D, E): resource system, resource units, actors, governance systems, and external environment (Ostrom 2009). The second level are twenty-four factors (identified in Section 4.2.3 as 1., 2., 3. etc.). They are a synthesis of existing success factor syntheses (for their derivation see Section 5.1) and are therefore based on a broad empirical data basis that has gone through a peer review. The third level consists of 54 criteria which break down the success factors in detail (1.1, 2.3, 4.2, etc.). These criteria in turn consist of 120 indicators which, in contrast to the success factors or criteria, can be measured directly (labels a, b, c). Ecological success is also measured using a total criteria of 4 and associated indicators (a total of 8). It has been pointed out that several ecological indicators are superior to just one (Birkhofer et al. 2015).

The indicator system itself must be validated. In this case, this was done through expert interviews. Some of the experts were interviewed (Prof. Dr. Elinor Ostrom, Prof. Dr. Michael McGinnis), some asked via an online survey. This was realized with the open-source software Limesurvey, version 1.90 (http://www.limesurvey.org). Seven experts took part in the survey. They were selected from the Workshop for Political Theory and Policy Analysis and the 2010 SES Club conference in Bloomington, Indiana, USA. Due to time constraints, the experts could not be expected to evaluate all 120 indicators. Instead, the 54 criteria (one level up) were queried, but the 120 associated indicators were mentioned in brackets. The time required for the survey was about 30 minutes.

The survey tested (a) the quality of each sub-concept (criterion), (b) whether the respective sub-concepts covered the corresponding success factor, and (c) the importance of each sub-concept.

Each question about the quality of the criteria ("How do you judge the quality of this criterion?") was a Likert scale ("Very good" = 1, "Good" = 2, "Medium" = 3, "Bad" = 4, "Very bad" = 5, or "Do not know"). A definition of quality was given for each question (quality = the criterion measures what it should and is reliable). This is not discussed further here.

For each group of criteria, that is, the sub-concepts of each success factor, *coverage* was tested, that is, whether important aspects were missing. This step ensured that the relevant factor was covered as fully as possible. We checked the coverage by asking "What percentage of the success factor is covered?" If the answer was lower than 65% coverage, a follow-up question popped up: "What criteria do you miss for this success factor?" This answer could be entered as open text in a comment field. As result, we find an average coverage of a factor of 77.2%.

For weighting purposes, the *relative importance* of each criterion in direct comparison to the other criteria for a factor was also asked (see Figure 4.6). This is done on a scale of 1 to 10, with 1 corresponding to a 10% weighting and all criteria together yielding 100%. This result was used for deleting sub-concepts that were considered unimportant by all experts. However, all sub-concepts were considered relevant.

Finally, we asked the experts how *relevant* each sub-concept was for ecological success. Table 4.2 shows the corresponding evaluation of this expert survey.

It is noticeable that the grades are relatively good throughout. The overall average is 1.94. The worst criterion is the grade 2.5 for "Adaptability to changes by nature." The best average grade is 1.43, which is the criterion for "dependency on resource."

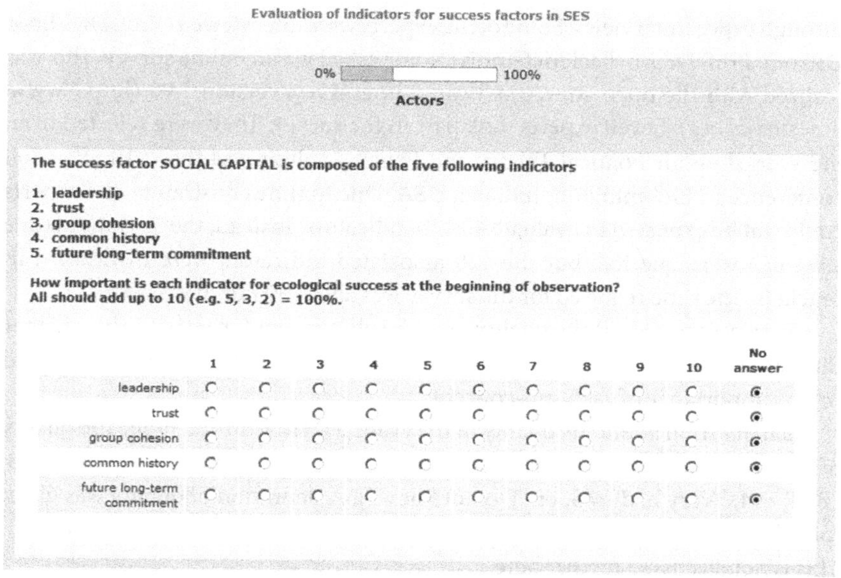

Figure 4.6 Screenshot of the expert survey to assess the quality of criteria (named indicators here) for each of the twenty-four success factors

Table 4.2 External evaluation of the relevance of the success factors

Number success factor	Acronym	Name of success factor	Criterion	Average evaluation of relevance
Resource system				
1	F1	Resource size	Quality of I–1: area	2.38
2	F2	Resource boundaries	Quality of I–1: boundaries	1.75
3	F3	Accessibility	Quality of I–1: location	2.00
3	F3	Accessibility	Quality of I–2: barriers	2.00
4	F4	Ecological success at the beginning	Quality of I–1: appropriation	2.00
4	F4	Ecological success at the beginning	Quality of I–2: maintenance	2.25
4	F4	Ecological success at the beginning	Quality of I–3: externalities	2.00
Resource units				
5	F5	Manageability	Quality of I–1: mobility	2.00
5	F5	Manageability	Quality of I–2: predictability	2.13
5	F5	Manageability	Quality of I–3: storage possibilities	1.88
5	F5	Manageability	Quality of I–4: ease of harvest: distribution of RU	2.13
5	F5	Manageability	Quality of I–5: ease of harvest: ease of finding RU	1.71
6	F6	Regeneration of RU	Quality of I–1: time of regeneration	1.75
Actors				
7	F7	Number of actors	Quality of I–1: number of actors	1.43
8	F8	Group composition	Quality of I–1: group composition	1.71
9	F9	Social capital	Quality of I–1: trust	2.00
9	F9	Social capital	Quality of I–2: group cohesion	2.14

Continued

Table 4.2 *Continued*

Number success factor	Acronym	Name of success factor	Criterion	Average evaluation of relevance
9	F9	Social capital	Quality of I–3: long–term commitment	2.00
9	F9	Social capital	Quality of I–4: common history	2.00
10	F10	Dependency on resource	Quality of I–1: dependency on resource*	1.43
10	F10	Dependency on resource	Quality of I–2: utility	2.29
11	F11	Dependency on group	Quality of I–1: dependency on group*	1.86
11	F11	Dependency on group	Quality of I–2: willingness to invest	2.14
Governance system				
12	F12	Group boundaries	Quality of I–1: group boundaries	1.57
13	F13	Participation	Quality of I–1: institutions	1.86
13	F13	Participation	Quality of I–2: rights	2.14
13	F13	Participation	Quality of I–3: possibilities to communicate	2.14
13	F13	Participation	Quality of I–4: leadership	2.00
14	F14	Legal certainty and legitimacy	Quality of I–1: rights	1.86
14	F14	Legal certainty and legitimacy	Quality of I–2: stability	1.86
14	F14	Legal certainty and legitimacy	Quality of I–3: recognition by others	1.86
15	F15	Administration	not asked, later success factor	—
16	F16	Information	Quality of I–1: communication	2.00
16	F16	Information	Quality of I–2: information	1.86
17	F17	Characteristics of rules	Quality of I–1: ease of understanding	2.00

Table 4.2 *Continued*

Number success factor	Acronym	Name of success factor	Criterion	Average evaluation of relevance
17	F17	Characteristics of rules	Quality of I–2: flexibility	1.71
17	F17	Characteristics of rules	Quality of I–3: feedback system	1.86
18	F18	Fairness	Quality of I–1: equality	2.00
19	F19	Control	Quality of I–1: control	1.71
20	F20	Compliance	Quality of I–1: following of rules	1.71
20	F20	Compliance	Quality of I–2: enforcement	2.00
20	F20	Compliance	Quality of I–3: sanctions	2.14
21	F21	Conflict management	Quality of I–1: type of conflict	2.29
21	F21	Conflict management	Quality of I–2: local jurisdiction	1.57
21	F21	Conflict management	Quality of I–3: mediation through third parties	1.86
External Environment				
22	F22	Exclusion	Quality of I–1: rights	2.00
22	F22	Exclusion	Quality of I–2: possibilities	1.86
22	F22	Exclusion	Quality of I–3: breaking of rules	2.33
23	F23	Relations	Quality of I–1: collaboration	1.80
23	F23	Relations	Quality of I–2: conflicts	1.60
24	F24	Capabilities to adapt to change	Quality of I–1: nature	2.50
24	F24	Capabilities to adapt to change	Quality of I–2: markets	2.33
24	F24	Capabilities to adapt to change	Quality of I–3: technologies	2.00

Continued

Table 4.2 *Continued*

Number success factor	Acronym	Name of success factor	Criterion	Average evaluation of relevance
Ecological success				
25	O1	Ecological success	Quality of I–1: stability	1.71
25	O1	Ecological success	Quality of I–2: condition of resource	1.71
25	O1	Ecological success	Quality of I–3: externalities	2.14

Note: Factors 10 and 11 were later renamed; the later name is already used here; therefore, indicator and factor are named the same here (each I-1); grading is like the school system (1 = best, 6 = worst)

Together with the good coverage, the good relevance throughout, and the consistently positive assessment in the oral interviews, this results in a satisfactory validation of the indicator system.

4.2.3 Overview about the indicators used

In order to be able to work out the two upper levels just mentioned, namely factors and criteria, it was necessary to start from a finer categorization system. I call this level, which is below the criteria the indicator level. The following list presents an overview of the indicators. The highest level are the five categories, denoted by capital letters, followed by the twenty-four success factors, denoted by numbers, which again have subsections, called criteria, denoted by numbers with decimals (e.g., 1.1, 2.4, 3.2, . . .). The lowest level are indicators, denoted by lower case letters. The indicators are the most detailed conceptual breakdown in this analysis. They are an attempt to cover the criteria as completely as possible. They are an additional aid in assigning the variables to factors and criteria.

A Resource
  1 *Size of the resource*
      1. Area of the resource system
          (a) Area in m^2
          (b) Time needed to pass through the system by usual means

  2 *Boundaries of the resource*
      1. Clarity of boundaries

   (a) Spatial clarity (number of overlaps)

   (b) Users' knowledge of the limits of the resource

  3 *Accessibility*

   1. Location of the resource

    (a) Time needed from place of residence to resource

   2. Physical barriers

    (a) Length of preparation time before appropriation can begin

    (b) Number of appropriation restrictions for a given time slot

  4 *Ecological success at the beginning of the observation period*

   1. Stability

    (a) Duration of the existence of the SES

   2. Quality of resource system

    (a) Condition of maintenance

    (b) Degree of biodiversity

   3. Quality of resource units

    (a) Degree of intensity of use

    (b) Balance between removal and regeneration

   4. Externalities

    (a) Degree of degradation

    (b) Level of pollution

    (c) Number of technical externalities

B Resource units (RU)

  5 *Manageability*

   1. Predictability

    (a) Degree of how easy the RU is to find

    (b) Degree of complexity and dynamics of the system

    (c) Existence of information on resource units

   2. Degree of harvesting process difficulty

    (a) Mobility of the RU

    (b) Accessibility

    (c) Degree of appropriation difficulty

   3. Processing

    (a) Means of transport

    (b) Storage possibilities

  6 *Regenerative capacity of the resource units*

   1. Pressure on resource units

C Actors

  7 *Number of actors*

   1. Number of users

    (a) Number of subgroups involved

    (b) Number of users

8 *Group composition*
   1. Homogeneity of the group with respect to . . .
     (a) Hierarchies
     (b) Interests
     (c) Socio-cultural identity
     (d) Gender distribution
     (e) Educational level
     (f) Wealth

9 *Social capital*
   1. Leadership
     (a) availability
     (b) locality
     (c) experience
     (d) reliability
     (e) same interests as the user group
   2. Trust
     (a) Degree of confidence
   3. Group cohesion
     (a) Way of interaction (e.g. with regard to reciprocity)
     (b) Degree of user interconnectedness
   4. Common past
     (a) Duration of traditions
     (b) Number and strength of common moral standards
     (c) Community experience of CPR and/or collective action
   5. Future, long-term commitment / voluntary commitment
     (a) Passing on knowledge, traditions, values to next generation

10 *Dependency on resource*
   1. Property and investments
     (a) Percentage of actors owning the resource
     (b) Percentage of landowners
     (c) Level of investment in the resource
   2. Income and benefits
     (a) Amount of income
   3. Alternatives
     (a) Number of alternative sources of resource units
     (b) Amount of alternative income
     (c) Possibility of insurance against crop failures

11 *Dependency on group*
   1. Joint activities of group
     (a) Degree of dependence on joint activities
   2. Alternatives
     (a) Presence and number of other recipient groups

D Governance systems

    12 *Boundaries of the group*
        1. Boundaries of the group
            (a) Number of groups and presence of boundary criteria (structural clarity)
            (b) Users' knowledge of group boundaries

    13 *User participation*
        1. Collective rights
            (a) Existence of rights to change control systems (constitutional choice rules)
        2. Institutions
            (a) Degree of democracy (the majority of users are involved)
            (b) Number of nesting of organizational levels
        3. Meeting places (arenas)
            (a) Presence and easy accessibility
            (b) Frequency of meetings
            (c) Proportion of users attending meetings
            (d) Number of users participating in elections

    14 *Legitimacy and legal certainty*
        1. Recognition of rights
            (a) Recognition of local jurisdiction
            (b) Recognition of local authority that creates rules
        2. Recognition by
            (a) State
            (b) NGOs and other organizations
            (c) Other recipient groups
        3. Existence of rights
            (a) Existence of property rights
            (b) Existence of other rights
            (c) Percentage of existing divisible rights
        4. Stability
            (a) Duration of the existence of the rule of law
            (b) Duration of important legal practices
            (c) Degree of corruption / bribery

    15 *Administration*
        1. Positions
            (a) Existence of rights of management
            (b) Legitimacy of the administration
            (c) Clarity of administrative positions

    16 *Information*
        1. Communication

      (a) Degree of ease

      (b) Level of effectiveness

      (c) Frequency and timeliness

    2. Information about

      (a) Resource (available / not available)

      (b) Resource units

      (c) User group

      (d) Control system

17 *Rules*

    1. Comprehensibility

      (a) Rules are written in the mother tongue of the majority of users

      (b) Degree of clarity and simplicity of rules

    2. Flexibility

      (a) Degree of adaptability of rules

    3. Feedback system

      (a) Existence of regular tests

      (b) Presence of tests

      (c) Existence of a feedback system for rule changes

18 *Fairness*

    1. Justice

      (a) Number and degree of inequalities between groups

      (b) Number of democratic characteristics of the institutions and rules

19 *Control*

    1. Monitoring

      (a) Number and effectiveness of biophysical monitoring

      (b) Number and effectiveness of social monitoring

20 *Compliance*

    1. Compliance

      (a) Degree of compliance

      (b) Degree of disregard

    2. Enforcement

      (a) Number of guards

    3. Penalties

      (a) Presence of a gradation mechanism

21 *Conflict mechanisms*

    1. Type of conflicts

      (a) Number of persons involved (proportional to the total number)

      (b) Degree of violence

    2. Local mediation

      (a) Presence and accessibility of local jurisdiction

(b) Characteristics of local jurisdiction

3. Mediation from outside

(a) Presence and accessibility of mediating institutions

E External environment

22 *Possibility of exclusion of third parties*

1. Rights

(a) Rights exist de jure

(b) Users know their rights

2. Implementation

(a) Actual possibilities of exclusion

3. Rule breaks

(a) Number of rule breaks

(b) Severity of rule breaks

(c) Possibilities of short-term profit for others

23 *Interactions with External Parties*

1. Cooperation

(a) With the state

(b) With organizations (e.g. NGOs)

(c) With other recipient groups

2. Conflicts

(a) With the state

(b) With organizations (e.g., NGOs)

(c) With other recipient groups

24. *Adaptability*

1. Nature

(a) Time of adjustment

(b) Number of changes resulting from the adjustment process

2. Markets

(a) Time to adapt

(b) Number of changes resulting from the adjustment process

3. Technologies

(a) Time to adapt

(b) Number of changes resulting from the adjustment process

F Ecological success

25 *Ecological success at the end of the observation period*

1. Stability

(a) Duration of existence of SES

2. Quality of resource system

(a) Condition of maintenance

(b) Degree of biodiversity

3. Quality of resource units

    (a) Degree of intensity of use
    (b) Balance between removal and regeneration
 4. Externalities
    (a) Degree of degradation
    (b) Level of pollution
    (c) Number of technical externalities

The indicators were extracted from the quoted studies during the systematic literature analysis (see Section 2.5). However, not all of them are fully operationalized. For example, "degree of biodiversity" needs to be further specified in order to make actual measurements, for example, via the Shannon index. This is not necessary for the analytical purposes of this work, as no new data is collected. Here the indicator level serves as the finest subdivision, which establishes the connection to the already existing variables through operationalization and allows allocation to success factors at a precise level. The indicator system is therefore not empirically validated as is customary in ecological studies. However, this was not the goal. Rather, the aim was to provide a comprehensible, transparent, and detailed description of the link between the top level (abstract concepts), the criteria, and operationalization (indicators and associated variables). In this step, all levels are displayed and described. The next step is to assign the existing variables to these theoretical concepts.

4.2.4 Assigning variables to indicators

Many studies point to the importance of certain factors in regard to ecological success in SES. From the analysis of these studies (see Sections 2.5 and 5.1), twenty-four potential success factors result as synthesis. In the previous section (Section 4.2.3), these factors have been subdivided even more precisely into sub-factors and indicators. The available information in variables must now be assigned to these indicators and thus automatically to the factors as well. The degree in which variables express an indicator, and the respective indicators their factor defines the model quality in the end.

The indicator system described in Section 4.2.3 guarantees a uniform allocation of variables to success factors, since associated criteria and indicators are already fixed before allocation. This assignment is the same for all data sets. Figure 4.7 shows this assignment process schematically without the criteria level.

Explanation: Variables (left) are selected (e.g., 1 and 2), combined (e.g., 4 and 5), or not used (e.g., 3) and finally summed to obtain values at the indicator level (middle). Indicators are in turn aggregated at factor level. Finally, these aggregations represent the data used for the analyses. The variables are assigned

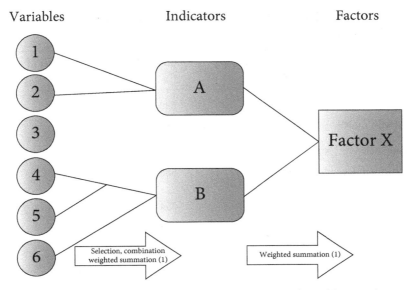

Figure 4.7 Conceptualization of selection and association of variables to indicators and factors

to the success factors by three independent raters. The assignments are compared and those that do not match are checked again independently and without discussion in a second round. This shows that for the CPR data set, for example, the satisfactory interrater reliability of 0.65 (Krippendorff's alpha) already rises to 0.77 by means of a brief comparison of understanding and already reaches a very good value of 0.85 in another independent second round of the examination of the disputed allocations (Table 4.3). Discrepancies that still existed were resolved in group discussions. The majority of the differences were due to misinterpretation of variables in terms of understanding and could be resolved quickly and uncontroversially. The interrater reliability of the three independent raters was good (Table 4.3, calculated according to Freelon 2013).

The final assignment of three independent raters for IFRI was again examined against the assignment of an IFRI forest expert. The level of agreement is still satisfactory. A higher agreement cannot be expected with this high number of categories (allocation to twenty-four factors + success). The interrater reliability, which may be estimated as good to very good, in conjunction with the well-developed indicator system means that the most serious error in operationalization is a certain degree of blurring, that is, that not all variables are included. However, even this risk is low, since all variables even with questionable relevance have been included; the rejected variables are *clearly* unusable.

Table 4.3 Krippendorff interrater reliability for assigning variables to factors for CPR, NIIS, and IFRI

Procedure	Krippendorff's Alpha	Number of cases	Number of decisions
CPR	0.65	588	1,764
CPR with short discussion of comprehension	0.77	588	1,764
CPR with independent second round	0.85	588	1,764
NIIS	0.77	554	1,662
IFRI	0.62	1,024	3,072
IFRI final assignment vs. forest expert	0.62	1,024	2,048

The technical implementation is carried out using a software shown in Figure 4.8. Each variable is read in individually, displayed with a description, and can be assigned to success factors via a tree structure. Figures 4.8 and 4.9 (partly German) illustrate the function of the assignment process.

In the green area in Figure 4.8 the variable name (red 1) is mentioned first. The abbreviation of three letters before the variable name (in this case "opl") identifies the category (here: operational rules) to which the variable belongs. The question itself (2) and possible answers (3) follow. If the variable cannot be assigned, there are further answer options below the marked area (4). If it is a good control question, the variable can be treated separately. If it can be used twice, two or more factors are assigned to it (label: "variable can be used twice"). If it is not clear how it should be assigned (label: "Don't know . . ."), it goes through the assignment process a second time. If it cannot be uniquely assigned this time either, its assignment is decided by the discussion of the three raters. If it is clear that the variable is not relevant for any of the success factors, it is sorted out as unusable (label: "Variable unusable"). Underneath are the top categories (5) and the success factors (6).

Figure 4.9 shows the actual assignment of a variable (*opl_arenfreq*) to one of the indicators (rights, institutions, arenas) of success factor 13 (participation of users).

This type of presentation reduces cognitive stress. This in turn avoids errors, as 120 indicators will always be available visually and do not have to be remembered. Moreover, transparency increases because the assignment possibilities are always presented visually and consistently. Combined with the fact that all raters use the same procedure, the assessment process is smooth.

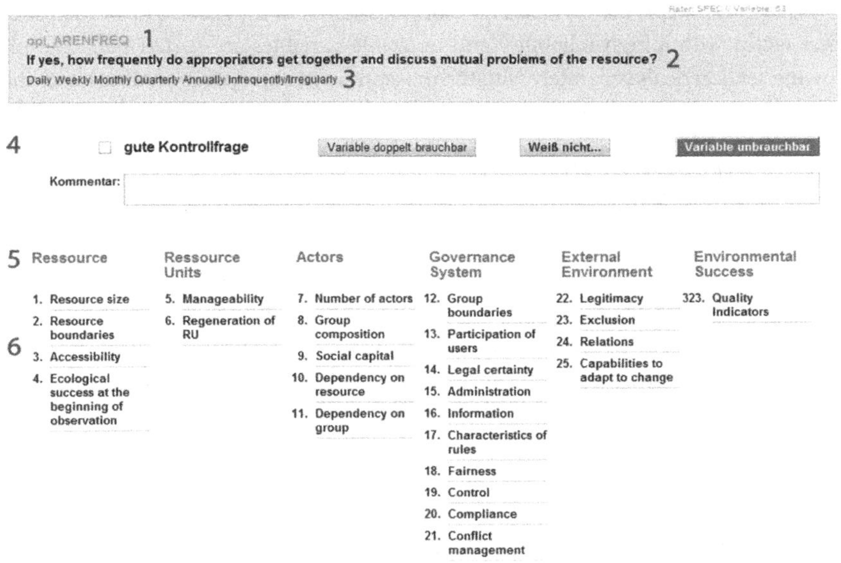

opt_ARENFREQ **1**

If yes, how frequently do appropriators get together and discuss mutual problems of the resource? 2

Daily Weekly Monthly Quarterly Annually Infrequently/Irregularly **3**

4 ☐ **gute Kontrollfrage**   [Variable doppelt brauchbar]   [Weiß nicht...]   [Variable unbrauchbar]

Kommentar: [   ]

5 Ressource

Ressource Units

Actors

Governance System

External Environment

Environmental Success

1. Resource size
2. Resource boundaries
6
3. Accessibility
4. Ecological success at the beginning of observation

5. Manageability
6. Regeneration of RU

7. Number of actors
8. Group composition
9. Social capital
10. Dependency on resource
11. Dependency on group

12. Group boundaries
13. Participation of users
14. Legal certainty
15. Administration
16. Information
17. Characteristics of rules
18. Fairness
19. Control
20. Compliance
21. Conflict management

22. Legitimacy
23. Exclusion
24. Relations
25. Capabilities to adapt to change

323. Quality Indicators

Figure 4.8 Assignment of variables to success factors—screenshot of software used

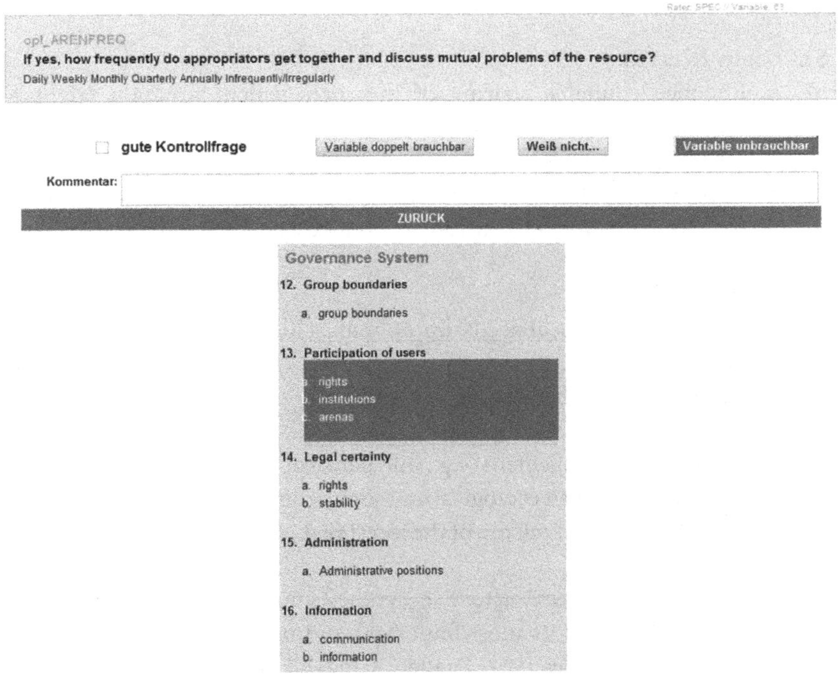

opt_ARENFREQ

If yes, how frequently do appropriators get together and discuss mutual problems of the resource?

Daily Weekly Monthly Quarterly Annually Infrequently/Irregularly

☐ **gute Kontrollfrage**   [Variable doppelt brauchbar]   [Weiß nicht...]   [Variable unbrauchbar]

Kommentar: [   ]

ZURÜCK

Governance System

12. **Group boundaries**

   a. group boundaries

13. **Participation of users**

   a. rights
   b. institutions
   c. arenas

14. **Legal certainty**

   a. rights
   b. stability

15. **Administration**

   a. Administrative positions

16. **Information**

   a. communication
   b. information

Figure 4.9 Selection of the appropriate success factor for a variable

After these steps, the variables have been assigned to the indicators of each success factor with a high reliability and unusable variables are sorted out. The following section deals separately with the measurement and operationalization of the dependent variable—ecological success—because of its significance for the model.

4.2.5 A difficult task—operationalizing ecological success (part 1—theory)

Success is usually divided into three aspects: ecological success (e.g., good condition of the resource), social success (e.g., equity), and economic success (e.g., efficiency or income). Positive evaluations in these three categories and their indicators are considered success. A widely cited example of conceptualizing success is a meta-analysis for forests (Pagdee, Kim, and Daugherty 2006):

S1. Ecological sustainability includes:
  A. Improve forest conditions (e.g., increase of forest area, species diversity, forest productivity, and number of valuable species)
  B. Address environmental degradation (e.g., reforestation, soil erosion protection, and watershed management)

S2. Equity refers to:
  A. Enhance equitable sharing of the management function (right to manage), entitlement (right to access and control), and responsibility for a given territory or set of natural resources
  B. Improve equitable benefit distribution among community members
  C. Increase investment in the future productivity of the forests

S3. Efficiency includes:
  A. Meet a range of local needs, improve local living standard, and alleviate poverty
  B. Reduce conflicts between local communities and authorities
  C. Control corruption
  D. Resolve mismanagement (e.g., imbalance of administrative power, and imbalance between ecological and socioeconomic dimension)
  E. Reduce individual misuse of the forest (e.g., timber smuggling)

Analogously to the success factors, a synthesis for success is carried out, involving restructuring of its most important and most frequently mentioned facets (Ostrom 1990; Berkes 1992; Pagdee, Kim, and Daugherty 2006). The following list presents its most important points:

1. Ecological performance/efficiency indicators
   - Condition (e.g., reduction or growth) of the resource
   - Sustainability
   - Productivity, resilience of the ecosystem
   - Biodiversity, number of valuable species
   - Dealing with environmental degradation (recognizing overexploitation, taking appropriate measures)
   - Stability
2. Social performance indicators
   - Justice (administration, access, appropriation, etc.)
   - Responsibility
   - Sustainability
   - Investments in future productivity
   - User satisfaction
   - Stability
3. Indicators of social efficiency
   - Sustainability
   - Investments in future productivity
   - Meeting local needs
   - Improving the local standard of living and reducing poverty
   - Conflict management
   - Degree of compliance
   - Balance between conflicting management objectives
   - The cost-benefit ratio of resource appropriation
4. Economic efficiency indicators
   - Productivity
   - Cost-benefit ratio of the appropriation
5. External effects on other SES
   - Ecological effects
   - Social effects
   - Economic effects

Concept 3, social efficiency, defines the degree to which actors are treated equally in terms of appropriation, participation in rule changes, etc. On an individual level, this means whether each individual feels they are being treated fairly. On the group level, success is whether the group has succeeded in achieving a Pareto-optimum. This means that no one can be better off without someone else being worse off. Concepts 2 and 3 (social performance and efficiency) can be summarized in one category if necessary. Almost no study focuses on concept 5 due to practical reasons, for example, unavailable data. A second reason is the frequent focus on institutions.

Apart from the question whether this conceptualizing does justice to defining success in a comprehensive way, it is not unproblematic to measure these indicators. Take productivity as an indicator of ecological performance: should we measure the productivity of a forest for a certain tree species, as biodiversity of multiple tree species, as trunk density, or density of growth? Each measurement has its pros and cons. A further complication is that the measurement of biodiversity is thus limited to a trophic group (here: trees), not taking into account biodiversity on other levels (Soliveres et al. 2016).

One proposal is to analyze several indicators using multivariate analysis (Wollenberg et al. 2007). However, precise data are usually not available, and so many studies are limited to one indicator, combined with the (subjective) assessment of foresters (or other specialists) or the data collectors themselves. This takes different forms—depending on resource and study—and is represented by *proxies*. For example, one study measures equity through equal access to bank loans and another measures equity through the allocation of equal amounts of water. Sustainability is illustrated in the same study via signs of overfishing combined with the satisfaction of fishermen about catches and catch composition (Berkes 1992). The following section (Section 4.2.6) describes how this analysis attempts to solve these problems and measures ecological success.

4.2.6 A difficult task—operationalizing ecological success (part 2—implementation)

Similar to the factors, ecological success is measured by indicators. These four indicators have already been described (see Section 4.2.3),

(a) stability of the system (duration of existence),
(b) quality of resource system,
(c) quality of resource units, and
(d) externalities produced.

Assigned to these indicators are a total of 33 variables for CPR, 79 for NIIS, and 69 for IFRI.

For several reasons I limit the target variable to the *ecological dimension* of success. *Economic success* could not be analyzed due to lack of data—data on income, value of resource units, or similar indicators were not available in any data set. For the output, however, high demands are made on measurability and data availability.

In contrast, *social success* contributes to the model as input. On the one hand, this can be justified methodologically, since predictive performance of classification algorithms usually decreases considerably when two or more target variables have to be classified (Reed and Marks 1999). This is because the combination of two dimensions makes patterns more difficult to identify.

On the other hand, this can be justified in terms of content. Social performance and social efficiency can be seen as preconditions for ecological success. These include, for example, social capital, which is the factor that enables a group to operate sustainably. Likewise, it is only possible to achieve long-term sustainability in a fair system, for example. Unfair control systems lead to noncompliance with rules by the disadvantaged and eventually to overexploitation (Ostrom 1990).

Ecological success is constructed differently in the three data sets. In CPR and NIIS (see Sections 3.1 and 3.2) success consists of the 33 (see Table 7.2) and 79 (see Table 7.3) variables mentioned. Some of these variables are partly qualitative assessments of local experts, most of them are quantitative data on harvest quality and water quantity. The success in IFRI (see Section 3.3 and Table 7.4), however, can be calculated even more robustly by using and comparing three different methods thanks to available data (Salk, Frey, and Rusch 2014).

These three methods are (a) a comparison to similar forests, (b) a comparison to "untouched" reference forests, and (c) expert assessment. Both basal area and biodiversity is taken into account. Surprisingly, all three methods are largely congruent, which has far-reaching consequences for future sustainability assessments in forests (Salk, Frey, and Rusch 2014). This is particularly important in regard to the enormous costs involved in accurately measuring the condition of forests, considering that in Germany alone, for example, in the Federal Forest Inventory 2011/2012, more than 420,000 trees were measured nationwide. Thus, existing data of expert assessment could replace planned forest inventories, since such assessment is the cheapest and fastest method by far.

The calculations of IFRI's ecological success are described in more detail in the following paragraphs (Salk, Frey, and Rusch 2014). Sustainability is measured by using two indicators, biodiversity and basal area of trees. Both were calculated in three different ways: (1) in absolute terms, (2) relative to nearby other IFRI forests, and (3) relative to independent reference forests that have remained virtually untouched and whose data came from other sources (Gentry 1988).

The base area of trees can be calculated from the available tree data. It is calculated separately per hectare for trees and saplings for each measured forest area. These areas are called plots. There are approximately 30 per forest, thus approximately 12,000 in total. The numbers of these basal areas have been added to the database for each plot. All species that are not trees have been excluded. The data for reference forests (Gentry 1988) had a different format and were adapted in a complex procedure.

In contrast to the base area, biodiversity is extremely difficult to assess or measure. This applies in particular to various tropical forest species. For forests, biodiversity is mostly measured per area or per tree. Since biodiversity per area is sensitive to stem density, which in turn depends strongly on age and regions, biodiversity per tree was measured. The difficulty of measuring biodiversity across different geographical regions (Sala et al. 2005), which have very different levels of biodiversity (Salk, Frey, and Rusch 2014), had to be taken into account. To compensate for differences in the number of plots measured, a species diversity median bootstrap procedure was performed for a sample of 100 trees (with resampling). One difficulty, for example, concerned the species names—duplicates were removed using a Java program written for this purpose and thousands of names were adjusted to avoid overestimating biodiversity.

Once both base area and biodiversity data were available for each IFRI forest, they were compared with nearby IFRI forests and reference forests. Base areas of most IFRI forests are similar and lie within two standard deviations to nearby forests. However, compared to the reference forests, basal areas of IFRI forests were lower throughout, with only some exceptions. This is not surprising, as the reference forests are older.

The diversity of IFRI forests and measurement techniques is also reflected in the diversity of species measured: the number of wood species ranges from 2 to 336. IFRI forests are mostly below the reference forests concerning biodiversity. Only 11 out of the 94 IFRI forests for which reference forest data were available had a higher biodiversity.

These two measuring techniques for assessing the forest condition (biodiversity and area) are largely independent of each other. This is again independent of whether scaled raw data or scaled data relative to the reference forests are used (Figure 4.10).

It can be concluded that measurement of the resource status for the IFRI data using three different methods—either assessment by experts, by local users, or by complex quantitative measurement—may be used almost equivalently (Salk, Frey, and Rusch 2014). The expert assessments are also available for CPR and NIIS and contribute to their output.

Hence, all pieces are now in place to present the results—the analytical methods (multivariate linear regressions, random forests, and artificial neural networks) were described in Section 4.1, the relevance of the success factors contributing to the model in Section 2.5, which is taken up again in Section 5.1. The synthesis of success factors has been linked to a newly developed indicator

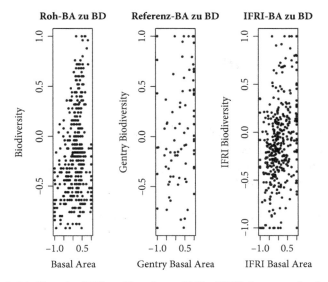

Figure 4.10 Biodiversity (BD) and basal area (BA) of IFRI-forests and reference forests in comparison

Note: The middle graph has fewer data points, because not all IFRI forests have a suitable reference forest.

system for SES in Section 4.2 and finally, the particular difficulties of measuring ecological success were presented in Section 4.2.6.

5

Results and Discussion

This chapter contains the results and the discussion. First, the development of the success factors synthesis is presented (Section 5.1). The contribution of each success factor is then critically discussed (Section 5.1.5). After this theoretical result, the main sections (5.2, 5.3, 5.4, 5.5, and 5.6) presents the modeling results for each data set, including descriptive statistics, correlations, and robustness and sensitivity analyses. Each result section is followed by a discussion.

5.1 Synthesis of success factors

In Section 2.5, previous research has been discussed which compiles a synthesis of factors that may be relevant for success. These experiments usually focus on identifying important system attributes for the analysis of SES. Between these attempts there are clear overlaps, sometimes even congruency. This is due partly to the research proximity to the "Ostrom school," partly to the obvious relevance of attributes such as size of the resource or number of actors. Additions such as market distance or the importance of social networks have often been included in later studies, so that there is a broad consensus as far as the majority of potentially relevant factors is concerned. However, this consensus is no longer given at the level of detail (e.g., meaning and exact formulation) or the type of categorization (e.g., location in the conceptual system).

An integration of these existing factor syntheses into one comprehensive synthesis, as presented in the following sections (Section 5.1.1–5.1.5), must therefore attach particular importance to three aspects: Firstly, completeness without redundancy. Secondly, logical consistency in terms of level selection, degree of abstraction of concepts, and membership of super-categories. Thirdly, a general but not misleading formulation of concepts at each level.

5.1.1 Justifying the choice of success factors

The three requirements—completeness, consistency, and generality—are tightly connected to operationalization. While the first point does not require any further explanation, the assignment of concepts to super-categories is often

Sustainable Governance of Natural Resources. Ulrich Frey, Oxford University Press (2020). © Oxford University Press.
DOI: 10.1093/oso/9780197502211.001.0001.

controversial. An example: Is a cow grazing a commons willow to be considered a "harvesting technology," or is it rather a resource unit and, as an animal, belongs to the natural resource system? Or is it both? While this may seem a contrived example to some, there are many fuzzy boundaries and unresolved questions like that throughout the aspired logically consistent conceptualization of an SES.

Another problem is the strong interdependencies of many concepts, such as between actors and the rule system. For example, actors who have reached a certain position can change the rules that legitimate their own position. In any analysis, it remains problematic to what extent the person should be separated from the position. In addition, many other questions remain.

A third problem is that SES are complex, extend across multiple sectors and long periods, and have different scales. With a two-dimensional, static list of factors such analytical challenges cannot be completely overcome. Nevertheless, such a synthesis of success factors is a first step towards a comprehensive SES analysis.

Finally, the question at which hierarchy level a concept should be placed cannot be answered in general, either, but depends on the research focus. For example, the size of an SES is usually of great interest and is a fundamental factor at the top of the hierarchy. In some systems, however, all cases may be of the same size—so this otherwise important parameter is no longer a distinguishing feature.

Heeding these requirements and tackling the problems mentioned, the twenty-four factors resulting from this synthesis represent independent concepts on a high and abstract level with minimal overlap. Since most syntheses neglect certain parts of SES, be they biophysical, economic, or social, this analysis uses several syntheses for the sake of completeness, while eliminating numerous overlaps.

Developing such a success factor synthesis may be driven by two different goals. On the one hand, a synthesis that is as comprehensive as possible can be targeted. Such an approach minimizes the likelihood of overlooking a factor and facilitates later analyses, since all factors that may be even remotely relevant are taken into account. Selecting factors of interest is then possible at a later stage, depending on the research objective. The obvious disadvantage is the consideration of many factors that ultimately do not play a role at all. Furthermore, it is fairly hard to construct a manageable model from such a multitude of factors. Moreover, there are unlikely to be any data for many factors.

On the other hand, the goal may be a minimum synthesis, which would take into account only those factors that certainly play an important role. Such an approach makes a meaningful model possible. However, it always runs the risk of ignoring a few factors in individual case studies. For reasons of research practice, however, such a minimum model often makes sense and is useful. Both approaches are presented in Section 5.1.2 and 5.1.3.

5.1.2 A comprehensive synthesis (which is too unwieldy)

The comprehensive synthesis is the result of the combination of success factors from the syntheses described in detail in Section 2.5 (Agrawal 2001; Pomeroy, Katon, and Harkes 2001; Pagdee, Kim, and Daugherty 2006; Schurr 2006; Gruber 2008; Shiferaw, Kebede, and Reddy 2008; Gutiérrez, Hilborn, and Defeo 2011; Cinner et al. 2012; Brooks, Waylen, and Borgerhoff Mulder 2012). It consists of 260 concepts and covers five levels as well as a large part of the concepts considered relevant in the literature on SES. These ten syntheses (see Section 2.2.5) are again based on many case studies or are often meta-analyses themselves. The factors are subordinated to the four main components of the SES framework model. My synthesis covers all factors and concepts found in the syntheses discussed. Intermediate hierarchy levels have been introduced, similar concepts have been grouped under one name and duplications have been deleted (definitions can be found in Section 5.1.5). Due to its length, the comprehensive synthesis can be found in the Appendix, Section 7.5.1.

Special attention is paid to selecting the most suitable levels for the respective concepts. Since this is done according to several criteria, such as abstractness, scope (number of sub-concepts), and theoretical and practical relevance for SES, there is no right or wrong here. For example, "monitoring" contains a very small number of subordinate concepts and would therefore have to be classified as low, that is, relatively unimportant. However, it is one of the central and best-studied concepts in SES literature (e.g., van Laerhoven 2010; Gibson, Williams, and Ostrom 2005; Chhatre and Agrawal 2008) and is therefore assigned a comparatively high level.

However, due to its size, this synthesis is difficult to handle for empirical studies. Due to the high number of sub-concepts, no data is available for the majority of its sub-concepts, even in extensive data surveys. This is a serious shortcoming. In addition, there are overlaps. This comprehensive synthesis is therefore particularly useful as a theoretical structure, for example for the design of data surveys (questionnaires) or for the creation of databases. It is too detailed for practical use with existing data and will not be discussed further here. Instead, the most important concepts are recombined into a minimum synthesis.

5.1.3 A minimal synthesis (which is about right)

The following minimum synthesis is suitable for analysis in several respects. On the one hand, the most important theoretical concepts are covered, on the other hand, it is sufficiently manageable for (field) work with data, that is, it is practical.

Furthermore, the databases used here (CPR, NIIS, IFRI) contain sufficient data for each of these twenty-four concepts. If variables are aggregated, meaningful statements about each of these factors can be made. This is already a considerable extension of all previous analyses with regard to number and width of covered variables. The synthesis looks is presented below (definitions and explanations of the factors can be found in Section 5.1.5):

Resource
1. Resource size
2. Resource boundaries
3. Accessibility
4. Ecological success at the beginning of observation

Resource units
5. Manageability
6. Regeneration of resource units

Actors
7. Number of actors
8. Group composition
9. Social capital
10. Dependency on resource
11. Dependency on group

Rule system
12. Group boundaries and exclusion rights
13. Participation of users
14. Legal certainty and legitimacy
15. Administration
16. Information and communication
17. Characteristics of rules (local fit)
18. Fairness
19. Control
20. Compliance with rules
21. Conflict management

External environment
22. Exclusion of third parties
23. Relations to other stakeholders
24. Capabilities to adapt to change

Ecological success
1. Stability and longevity
2. Quality of resource system
3. Quality of resource units
4. Externalities

Since this minimal synthesis provides very broad and abstract concepts at the top level (unlike the SES framework which partly has variables instead of concepts as second tier), any branch can be subjected to a detailed subdivision according to analytical requirements. If a specific theoretical question requires a detailed analysis of one particular factor, this aspect can be particularly highlighted with the help of the comprehensive synthesis and the finely structured indicator system as well as a corresponding data collection (cf. Basurto, Gelchich, and Ostrom 2013).

The primary objectives of this synthesis were a comprehensive list of all relevant success factors for SES, a high theoretical consistency of concepts regarding their place and level in the hierarchy given their scope, and a high usability for analyses, that is, operationalizability. The integration of different syntheses allows an improved classification of concepts on the most suitable levels. In addition, logical errors in similar conceptualizations are easier to detect and correct.

As discussed in the preceding paragraphs, there is no right or wrong way to create such a synthesis. However, the question remains—do the selected factors really represent a viable synthesis? For this reason, these factors were submitted to experts in SES for evaluation and validation. All experts have empirical experience in their field (3 in irrigation systems, 1 in forest management, 1 in land use) as well as theoretical knowledge (1 professor, 2 doctors, 2 PhD students; all have acquired their academic titles in this research area).

The experts were asked to assess the relevance of each factor for environmental success in the management of common pool resources on a scale from 1 to 10 (with 10 as the grade for the greatest importance). The results are summarized in Figure 5.1.

The least important factor in Figure 5.1 (5.5) is group composition, and the most important factor is social capital (9.2). The overall average is 7.3 out of 10, which means that each expert's opinion is significantly above the arithmetic mean between important and unimportant (see Table 7.1 for the exact values). From this fact and further expert discussions, the synthesis of success factors can be considered sufficiently validated. This is particularly true since the actual, rigorous examination of the significance of the factors takes place via case study data (n = 794), that is, by assigning relevance through the models (see Sections 5.2, 5.3, 5.4, 5.5, and 5.6).

Finally, the question arises why an integration of the existing syntheses into a new synthesis was carried out at all and why an existing one was not used. There are several reasons for this: Although each of these syntheses represents an important theoretical development, none can be considered suitable for the present analysis. One reason is the frequently lacking degree of abstraction, that is, the lack of generality (e.g., factor synthesis 6 in Section 2.5.2.6). A second reason is the one-sidedness of the factors considered, for example the disregard of biophysical factors (e.g., factor syntheses 1, 2, 4, 5, 8, 9). A third is the specialization on a sector with partly very specific factors (e.g., factor synthesis 7 for

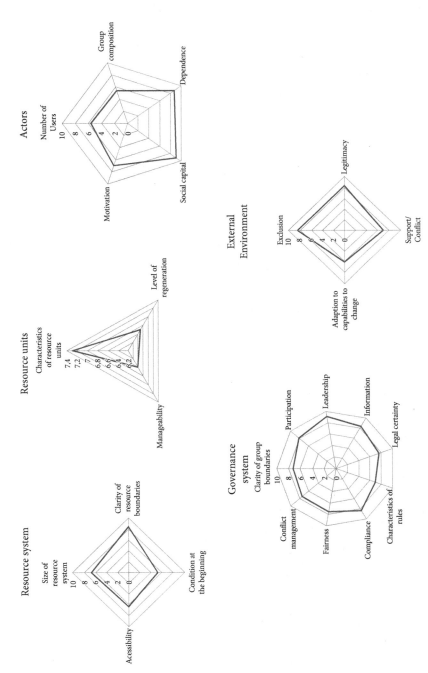

Figure 5.1 Mean value of relevance according to the assessment by five experts regarding the success factors for ecological success

fisheries). Only synthesis 3 and 10 meet the criteria of completeness, generality, and balance. They therefore serve as basis for my integrated synthesis, which tries to supplement missing concepts by using other sources. The justification for creating a new synthesis is the lack of a balanced and comprehensive system of SES-concepts. I will take up these points of criticism again in the discussion (Chapter 6) for the existing syntheses.

5.1.4 Which success factors are excluded?

Some factors are excluded from the analysis. One reason for exclusion concerns factors that are on a different level of abstraction. One example is "positive cost-benefit ratio" which is often cited as a success factor. However, this is a meta-factor: individuals intuitively calculate their potential gain by taking some or all of the above factors into account. Only if this rough calculation turns out to be positive, will they participate cooperatively in a joint management (Ostrom 1990).

Fuzzy concepts are another reason for exclusion. For example, some authors mention the "appropriate number of users to solve appropriation problems," which cannot be quantified at all since it is formulated too vaguely. Without expert knowledge in the respective case study, such "success factors" cannot be adequately quantified. Such criteria therefore cannot be taken into account in any comparative analysis.

A third reason for exclusion concerns concepts that are defined too broadly. One example is a "stable political, social, economic, and ecological environment." While certainly relevant to any SES, this "factor" is impossible to measure in this form due to its complexity. In addition, none of the case studies analyzed have systematic data for the respective SES environment.

This uniform theoretical analytical framework can now logically and consistently combine a large number of variables in a uniform language under suitable concepts. It is therefore also a guide for other researchers and has already been requested for this purpose and used for case studies, for example in African field studies and in English forest and park management (Aglionby 2014).

Before assigning data to these concepts, the success factors must be precisely defined. In addition, it must be explained in detail why each factor can potentially contribute to success. This is done in the next section (Section 5.1.5).

5.1.5 Why these success factors should be highly relevant

All success factors are explained in more detail in this section. In the past decades of social-ecological research, numerous factors have been named as possibly contributing to success. However, the actual relevance can only be proven by a quantitative analysis of many case studies, which has rarely been done (but see Gutiérrez, Hilborn, and Defeo 2011; or Brooks, Waylen, and Borgerhoff Mulder 2012). This analysis is the largest that I am aware of with a total of 794 case studies.

A synthesis of existing potential success factors has to demonstrate first—from a theoretical point of view—why and how these factors may contribute to success. These reasons are discussed in more detail on the basis of a literature review. As overview the most important contribution of each factor is summarized in Table 5.1, including literature references that provide a detailed justification.

Table 5.1 Relevance of success factors for success

Success factor	Relevance for success—influences . . .	Reference
F1 Resource size	Coordination and organization costs	Ostrom (2009)
F2 Resource boundaries	Transparency	Wade (1994)
F3 Accessibility	Appropriation costs	Nagendra (2007)
F4 Ecological success at the beginning	Status of resource and resource units	Ostrom (2009)
F5 Manageability	Appropriation costs	Ostrom (1992)
F6 Regeneration of resource units	Resilience against overexploitation	Baland and Platteau (1996)
F7 Number of actors	Coordination and organization costs	Olson (1968)
F8 Group composition	Coordination	Agrawal (2007)
F9 Social capital	Coordination	Gruber (2008)
F10 Dependency on resource	Willingness to invest	Agrawal (2007)
F11 Dependency on group	Willingness to invest	Pagdee, Kim, and Daugherty (2006)
F12 Group boundaries	Benefits benefit user groups, not third parties	Ostrom (1990)
F13 Participation	Efficiency	Ostrom (1990)
F14 Legal certainty and legitimacy	Willingness to invest	Schlager and Ostrom (1992)
F15 Administration	Organizational costs	Tang (1989)

Table 5.1 *Continued*

Success factor	Relevance for success—influences ...	Reference
F16 Information	Appropriation decisions	Sandström and Widmark (2007)
F17 Characteristics of rules	Efficiency	Meinzen–Dick (2007)
F18 Fairness	Willingness to invest	McKean (1992)
F19 Control	Defense against free riders	Ostrom (1990)
F20 Compliance	Defense against free riders	Pagdee, Kim, and Daugherty (2006)
F21 Conflict management	Coordination	Ostrom (1990)
F22 Exclusion	Appropriation decisions	Feeny (1992)
F23 Relations	Willingness to invest	Berkes (2007)
F24 Capabilities to adapt to change	Stability	Agrawal (2002)

In the following discussion of success factors, their relationships are given in brackets (for example F9). However, most of the time it is neither known nor statistically proven what these relationships consist of exactly, for example, in terms of cause and effect. A closer examination would go beyond the scope of this work; these largely unknown relationships are a major reason for the use of neural networks.

5.1.5.1 Resource system
F1—Resource Size

Definition/Description: The physical size of the resource system. It may be measured in area (e.g., forestry), length, and/or storage capacity (e.g., in irrigation systems).

What "size" ultimately means depends on the type of resource and the actors' perspective. Although there are some very large SES, in most cases size means no more than some hundred individual spots in fisheries or some hundred square kilometers in forestry. Moreover, depending on the means of transport and season (i.e., the accessibility, concept F3), the time to cross the system may vary considerably between two systems even if their size is the same.

Reasons, why important for success: If a resource is small, costs for a community to organize often exceed management benefits. If a resource is large, it is difficult and costly to collect knowledge (F16), monitor appropriation (F19), and control boundaries (Ostrom 2009). Many aspects of management are easier if there are clear arrangements (Wade 1992). The point of optimal appropriation is easier to find if the resource is small and well defined (F2), which in turn increases efficiency (Ostrom 1992a). Since resource size highly correlates with group size (F7), the larger the group becomes, the higher the costs to coordinate and the more difficult trust can be built (F9). Organization will be more difficult (Baland and Platteau 1996) and transaction costs will be higher (Shiferaw, Kebede, and Reddy 2008). It has been suggested that resource size does not affect success directly but through other pathways (Pagdee, Kim, and Daugherty 2006). Pomeroy, Katon, and Harkes (1998) indicate that resource size influences the way the resource is managed (e.g., management of individual species vs. ecosystem management). Size should also be in agreement with the number of people. Therefore, population density is of importance (Nagendra 2007).

Connections to other concepts: boundaries (F2), accessibility (F3), number of actors (F7), monitoring (F19)

F2—Resource boundaries

Definition/Description: The boundaries of a resource mark its geographical limits to the outside. This includes actors' knowledge of boundaries.

Reasons, why important for success: Clear boundaries facilitate initial efforts to cooperate and function as demarcation of externalities (Gruber 2008). Costs and benefits for appropriators are visible, which increases transparency (Gibson, Williams, and Ostrom 2005; Ostrom 1990). This is one of the original design principles—*clearly defined boundaries*—in "Governing the Commons" (Ostrom 1990). Without clear boundaries, protection against potential damage from the outside is difficult or impossible (Acheson 1987). Clear boundaries bring out the character of CPRs managed by a group, thus creating a higher motivation for mutual management (Pomeroy, Katon, and Harkes 1998; Wade 1992; Ostrom 1990).

Clear boundaries also facilitate the exclusion of third parties (F22, see Baland and Platteau 1996) and are a prerequisite for clear property rights (Schurr 2006) which in turn connects to other concepts (F14). For fisheries, a meta-analysis spells out these relationships:

> Spatial considerations, through clearly defined geographic boundaries (such as lake or enclosed bay) and sedentary life history of the resources contributed to

co-management success by confining the number of users, lowering associated costs of information gathering, monitoring and enforcement, and restricting the spatial dynamics of fishing effort to well-defined areas. (Gutiérrez, Hilborn, and Defeo 2011, p. 308)

In contrast, some authors argue for flexible and shiftable boundaries to adapt to different purposes of appropriators (e.g., Cleaver 2000). For forests, Pagdee, Kim, and Daugherty (2006) point out that congruence between the geographical and socio-economic boundaries is critical, whereas the existence of clear boundaries is not of paramount importance.

Connections to other concepts: resource size (F1), exclusion of third parties (F22), legal certainty (F14)

F3—Accessibility

Definition/Description: The ease with which resource system and resource units can be reached. Most important is the time needed to travel to the system and reach key locations. Accessibility includes both the distance from the actors' homes to the resource and the difficulty of extracting resource units. The extraction process can be constrained by geographical barriers or biophysical constraints (e.g., seasons).

Reasons, why important for success: If a resource system is large (F1), this may be of importance, because appropriation costs are lower if appropriators live nearby the resource, which in turn depends on the spatial and temporal distribution of resource units within (F5). A close distance also facilitates monitoring and thus improves compliance with rules (F19, F20, see Ostrom 1992a; Nagendra 2007; Baland and Platteau 1996). If the distance is small, markets sometimes are less important (Baland and Platteau 1996). The location of the actors' group in relation to other communities (villages) or stakeholders like NGOs (F23) is also relevant for actors' cooperation (Pagdee, Kim, and Daugherty 2006).

An easily accessible resource decreases the cost of resource unit extraction (Nagendra 2007). This may have negative effects, too (Ostrom 2009; Pagdee, Kim, and Daugherty 2006): if a resource is very easy to access, the exclusion of third parties may be difficult or impossible (F22), which in turn may increase the pressure on the resource units considerably (F6). It has been pointed out that the often-used indicator of "distance of system to the next road" can be misleading if officials live locally (Agrawal and Chhatre 2006).

Connections to other concepts: monitoring (F19), manageability (F5), regeneration (F6)

F4—Initial ecological condition

Definition/Description: The state of the resource system before an actors' group with a certain rule set begins to appropriate resource units from it. Ecological condition comprises the quality of resource system and of resource units as well as externalities and system stability.

Reasons, why important for success: Both extremes—a very good and a very bad condition—are equally detrimental to commons management. If resource units are available in abundance, there is no need to organize. If the resource is heavily exploited, there is little future prospect, which also prevents organization (Shiferaw, Kebede, and Reddy 2008). Moreover, rules are more difficult to establish. If an actors' group has experienced a certain lack of resource units, they understand better that cooperation is necessary to ensure sustainable management (Thomson, Feeny, and Oakerson 1992; Ostrom 2009). If, on the other hand, the resource is degraded too much, motivation to participate in management may be insufficient (Pagdee, Kim, and Daugherty 2006; Ostrom 2009). A counterexample is described by McKean (1992): due to excessive logging, Japanese forests suffered significant losses up to total degradation. In the light of this experience, strict control systems were created and sustainable management plans introduced.

A long-term perspective assessing the past, present, and future of the condition of the resource is essential, especially in slowly renewable resources such as forests (Nagendra 2007). As a result, the initial state of the resource is critical, because duration and scope of measures for sustainable management depend on it. Other studies stress that as well (Olsson, Folke, and Berkes 2004): central and essential information about systems can only be achieved through long-term monitoring which has to start with a reference point—the initial condition. Only on this basis, can a sudden deterioration of the system be detected and countermeasures be taken.

Connections to other concepts: dependency on resource (F10), information (F16), participation (F13)

5.1.5.2 Resource units
F5—Manageability

Definition/Description: Manageability describes the process of handling resource units. This includes how predictable the units are (availability of information; complexity and dynamics of system; ease of finding units) as well as the ease of harvesting (mobility and accessibility of units) and handling (storage and transportation).

Reasons, why important for success: All biological populations fluctuate in their abundance over time. The degree of fluctuation varies greatly among populations, affected both by external environmental or other factors that are either independent of or dependent on population density (May 1999). Predictability is especially important for resource flows (Pagdee, Kim, and Daugherty 2006). It may be hard for actors in a very unpredictable system to allocate resources or coordinate activities (Agrawal 2001). In that sense, this variable may be especially critical during the start-up phase of CPR management, since common-property regimes are generally found when "resources are scarce, dispersed, mobile, or variable in predictability" (Tucker, Randolph, and Castellanos 2007, p. 271). If the resource units' predictability is low, it becomes hard for communities to collect observations for a mental model of the system (F16, see Baland and Platteau 1996).

For many aspects of management, these basic characteristics of resource units are important. The unique distribution of units in time and space directly influences costs of appropriation (Pagdee, Kim, and Daugherty 2006). Extreme examples would be fish living stationary in a nearby bay and, in contrast, trees that have to be transported from remote and steep places (cf. Nagendra 2007). Moreover, a low mobility increases accuracy of prediction as to where the resource units actually may be found. This decreases search costs and in consequence costs of appropriation. Furthermore, costs for information about the location of the units decrease as well (Ostrom 2009). Appropriation costs are lower if appropriators live nearby the resource. This also facilitates monitoring (F19) and thus improves compliance with rules (F20, see Ostrom 1992a).

In addition, the location of the actors' group in relation to other communities (villages) is of relevance (Pagdee, Kim, and Daugherty 2006).

Second, storability is certainly an issue as well (Agrawal 2001). Resource units that are easily storable and do not deteriorate rapidly, can be managed more flexibly regarding consumption, use, and sale. As a result, inevitable perturbations can be balanced better, which in turn stabilizes the system. In addition, rules are followed more closely since the immediate pressure to sell is lower (Agrawal 2001). This fact becomes particularly salient in times of need if the appropriated amount would damage the resource irreparably, but is necessary for the survival of the actors. Examples for highly mobile units that are problematic to store are fish; trees are an example for immobile units that are easy to store.

Connections to other concepts: accessibility (F3), information (F16), monitoring (F19)

F6—Regeneration of resource units

Definition/Description: The growth or replacement rate describes how fast extracted resource units replace themselves (Dietz et al. 2002). This allows calculating the pressure on them, if the extraction rate is given. Key concepts include the regeneration time between harvests (life cycle of units)—that is, the time until resource units have been replenished completely (e.g., water) or have reached maturity (e.g., fish)—and the amount that may be extracted without degrading the resource (maximum sustainable yield). The latter constitutes a key parameter influencing many other critical concepts (Pagdee, Kim, and Daugherty 2006).

Reasons, why important for success: The higher the regeneration rate of resource units, the more units may be appropriated without degrading the resource system (Baland and Platteau 1996). One important aspect is time between harvests, that is, until resource units have been replenished completely (e.g., water) or have reached maturity/adult size (e.g., fish or trees). Appropriated amount and time of regeneration cannot be separated sensibly. It is, for example, perfectly possible to harvest sustainably even great yields if the resource units regenerate quickly.

If, on the other hand, regeneration capability is slow, even a low rate of use may not be sustainable (cf. Agrawal 2001). Thus, regeneration time and capability have a direct and large effect on the condition of the resource (F4, see Wade 1992).

Regeneration, in turn, is itself influenced by many concepts. Chhatre and Agrawal (2008) suggest that rule enforcement (F20) is important with lacking enforcement leading to degradation. Group size (F7)—at least in forestry—seems to have a curvilinear effect on regeneration rates (Nagendra 2007): both too few and too many individuals per hectare of forest are not optimal for regeneration.

Baland and Platteau (1996) distinguish between two situations: a sustainable steady state versus an unstable equilibrium accompanied by degradation. In most systems, it is extremely difficult to find a steady state equilibrium because system dynamics are notoriously hard to predict (F5) and knowledge about the SES (F16) is often poor (Schlager, Blomquist, and Tang 1994). Therefore, actors are sometimes not aware that their restraint in the present (preserving the equilibrium) means higher yields in the future. It is even harder to preserve an equilibrium if only a few people are willing to invest labor into the SES (Baland and Platteau 1996).

Connections to other concepts: information (F16), condition at the beginning (F4), number of actors (A1)

5.1.5.3 Actors
F7—Number of actors

Definition/Description: This refers to the number of actors in the actors' group who are appropriators.

Reasons, why important for success: The larger the group, the higher the costs for coordination, organization, and implementing rules and activities (Olson 1968).

In addition, large groups increase the pressure on resource units (F6) by their very number (Baland and Platteau 1996; Agrawal and Chhatre 2006). In some cases though, costs for management and monitoring (F19) are higher for smaller groups (Ostrom 2009). Up to now, however, to our knowledge no study could establish a robust causal connection between number of actors and ecological success (cf. Pagdee, Kim, and Daugherty 2006). This may be due to different levels of heterogeneity within the actors' group (F8), which may have a substantial influence on that. Other studies (e.g., Agrawal and Yadama 1997) are also unable to find a clear relation and suggest a non-linear relationship. Chhatre and Agrawal (2008) do not find a (negative) connection between group size and condition of resource if type of management is controlled for.

Nevertheless, the size of the actors' group is of definitive relevance (Berkes 1986; Berkes 1992; Feeny 1992): trust (F9) builds up more easily in smaller groups, and rule violations are detected more easily (F20), especially third parties without rights to appropriate (F22). Apparently, there is a connection between group size and tenure regime of forests, too: leasehold forest users depended the most on the forest and had the smallest group size (Nagendra 2007).

Connections to other concepts: regeneration (F6), social capital (F9), group composition (F8)

F8—Group composition

Definition/Description: This factor describes the level of heterogeneity in the actors' group regarding different sub-concepts like ethnicity, wealth, education, gender, etc.

Reasons, why important for success: In general (with known exceptions), a high degree of heterogeneity (in race, ethnic group, class, clan, wealth, etc.) weakens group cohesion. Heterogeneity may also exist in regard to diverging interests. This influences coordination negatively. High heterogeneity may also impede fair distributions and collective action (Gruber 2008).

Since heterogeneity may refer to many aspects like ethnic group, culture, gender, wealth, or interests (Baland and Platteau 1996), it is difficult to subsume it under one concept, because while heterogeneity in one aspect may be positive, it may well be negative in another. Therefore, group composition is a mixture of highly context-dependent variables (Agrawal 2007). Still, some authors suggest that particularly in social aspects, high heterogeneity is negative in the majority of cases (Agrawal and Chhatre 2006; Tang 1992; Dayton-Johnson 2000). Other authors see no clear empirical relationship, because too many variables interfere (e. g. Pagdee, Kim, and Daugherty 2006; Pomeroy, Katon, and Harkes 1998). In a similar vein, other studies see no systematic connection between high heterogeneity and ecological success (Baland and Platteau 1996, p. 302ff). However, they suggest that high heterogeneity may affect the capacity to communicate effectively (F16); other negative consequences are mentioned as well.

This conclusion is reinforced by the fact that rules are easier to find in relatively homogeneous groups (Berkes 1986) and that group cohesion is better (Pomeroy, Katon, and Harkes 1998), which is a prerequisite for long-term planning and organization (Ostrom 2009). One study about heterogeneity sums this up as follows:

> Heterogeneity is not a strong predictor of the level of collective activity. Rather, heterogeneity is a challenge that can be overcome by good institutional design when the interests of those controlling collective-choice mechanisms are benefited by investing time and effort to craft better rules. (Varughese and Ostrom 2001, p. 747)

Connections to other concepts: social capital (F9), information (F16), fairness (F18)

F9—Social capital

Definition/Description: Social capital may be defined as robust local social networks, established norms and trust between the members of a particular community facilitating coordination, and cooperation (Pretty 2003; Gruber 2008; Schurr 2006). This is just one of many definitions that have been suggested.

Reasons, why important for success: Trust is perhaps the most essential part of social capital. It may be increased by already established norms and past experiences of groups with collective action. Building trust takes time (Olsson, Folke, and Berkes 2004) and works best when groups are not too large (F7). In turn, trust is essential for working together successfully over longer periods of time. Trust may be established between persons, but it extends to processes and institutions as well; high trust reduces transactions costs and furthers the willingness to invest in the system (Schurr 2006). In general, high trust means less monitoring (F19) and a higher compliance with rules (F20), which is a condition for sustainability (Gibson, Williams, and Ostrom 2005). Especially transactions based on reciprocity like the exchange of goods become easier and cheaper; contracts and arrangements are less complicated and less prone to defection (Ostrom 2009). Group cohesion increases with trust as well (Gutiérrez, Hilborn, and Defeo 2011). Therefore, the loss of reputation can be an important part of sanctioning in a well-functioning society (Wade 1992).

A common history of collective action may be a valuable reference point for tasks in the present (Baland and Platteau 1996), which is often the precondition for the motivation to plan, act, and invest for the long term (*long-term commitment*, see Scheberle 2000). Future long-term commitment includes transfer of knowledge, traditions, and values to the next generation. A common *history* and experience with collective action arrangements is particular advantageous in order to develop adapted, just, and effective rules (F17, see Gutiérrez, Hilborn, and Defeo 2011). Additionally, Pagdee, Kim, and Daugherty (2006) indicate that existing traditions and practices decrease the relevance of markets, because the community is, in comparison, more important. It is not clear whether there is a connection between social capital and tenure regime in forests (Nagendra 2007).

Social capital typically increases if arenas for collective action are made available. Another significant aspect is good leadership (Gutiérrez, Hilborn, and Defeo 2011; Scheberle 2000). Leaders should not be corrupt, should have the majority's support, and should have a high level of credibility. Ideally, good leaders are able to adapt to new situations, react flexibly, and learn from experience (Gruber 2008). Case studies have shown repeatedly that a dysfunctional system often results if an important leader discontinues his or her tasks (Pomeroy, Katon, and Harkes 1998).

Connections to other concepts: relations (F23), number of actors (F7), adapted rules (F17)

F10—Dependency on resource

Definition/Description: This concept indicates the degree of dependency on resource for survival. The subsistence level is influenced by the availability of alternatives, by ownership, and by the benefits that can be extracted from the resource.

Reasons, why important for success: Appropriators who depend heavily on the resource because they have no alternatives are more likely to invest in the resource system and to manage it sustainably. This is particularly true for dependencies that extend over several generations. This has been found both in field studies and laboratory simulations (Bischoff 2007). Resources that are critical for subsistence are, on average, better maintained because of their inherent high value (Ostrom 2009). It is not clear whether this dependency is only positive. For example, in the case of high poverty a strong dependency may lead to overuse or destruction of the resource (F6, see Agrawal 2007). In some studies, no link is found between dependency and success (here: forest condition, see Pagdee, Kim, and Daugherty 2006).

Nevertheless, a high dependency is often one of the keys for the initiation of self-organization (Pagdee, Kim, and Daugherty 2006; Gruber 2008; Ostrom 2009; Agrawal and Chhatre 2006). There is also a correlation between tenure regime and dependency (Nagendra 2007). This is a consequence of the fact that a high dependency necessarily means being dependent on appropriated resource units. As a result, one of the main reasons individuals join communities is their desire to minimize risks, because often only a community can buffer high and unpredictable risks like crop failures (Wade 1992). Chhatre and Agrawal (2008) point out that dependency is mediated by various other institutional factors. Therefore, it is not surprising for them that studies suggesting a direct link come to contradictory results. Furthermore, they distinguish between two forms of dependency: subsistence dependency and economic dependency. Lastly, whether there is only a strong concern about the resource's continuing existence or a real high dependency, the effect will be the same (Scheberle 2000).

Connections to other concepts: regeneration (F6), accessibility (F3), participation (F13)

F11—Dependency on group

Definition/Description: The group members are dependent upon each other regarding provisioning, distribution, and/or appropriation of resource units (Baland and Platteau 1996).

Reasons, why important for success: Group formation often starts with the insight or compulsion that successful and sustainable operations are only possible in groups (McKean 1992). Therefore, dependency on the group is often crucial for the initiation of a community-wide system of rules and regulations (Ostrom 1992a). This may also mean that exploitation of a resource by a single individual is not possible or that a group has to act as a unit to enforce their rights against outside parties. A strong dependency among group members can also be a prerequisite for collective actions to prevent third parties from entering the system (Berkes 1986). A high dependency may be bound not only by a common purpose but also by geographical isolation, strong cultural traditions, or similar factors.

If dependency on other group members is high, members are anxious to maintain reciprocity and trust. This contributes to a long-term view of actors on resource exploitation and increases the willingness to invest in the resource (F10) as well as social sustainability (Pagdee, Kim, and Daugherty 2006). If the dependency on resource is high (F10), the dependency on the *group* is often high as well, since strong fluctuations in the flow of resource units or losses from disasters can only be overcome together (Wade 1992).

Connections to other concepts: dependency on resource (F10), participation (F13), social capital (F9)

5.1.5.4 Governance systems
F12—Group boundaries

Definition/Description: This factor describes who belongs to the actors' group and who does not. This includes the knowledge about group boundaries of the actors themselves.

Reasons, why important for success: A meta-analysis (Cox, Arnold, and Villamayor Tomas 2010) of design principles (Ostrom 1990) demonstrates that many authors distinguish between clear boundaries of the resource system and clear boundaries of the actors' group. Group membership is defined by *boundary rules*. They can be implemented in many different ways. In irrigation systems, membership is often based on land ownership, interest to water rights, or payment of a fee but can be in fact any other socio-economic attribute (F8, see Tang 1991). Group boundaries are, on average, clearer in small groups (F7) with a common history of use (F9). They become especially important with highly mobile or scattered units (F5).

Clear boundaries of the appropriating group are an essential prerequisite for reaching decisions or agreements on a common system of rules. Clear boundaries facilitate the demarcation of externalities and costs while benefits for the appropriators become calculable. Furthermore, clear boundaries facilitate the exclusion of third parties (F22). Exclusion makes sure that appropriated units benefit exclusively the actors' group (Ostrom 1990). In addition, clear boundaries facilitate monitoring of compliance with rules (F19, F20), since rights can be assigned clearly (Pomeroy, Katon, and Harkes 1998; Schlager and Ostrom 1992).

Connections to other concepts: number of actors (F7), monitoring (F19), group composition (F8)

F13—Participation of actors

Definition/Description: The level of involvement of actors in collective action situations. Existing rights, arenas, and institutions are important aspects of participation.

Reasons, why important for success: This certainly is the one concept that is most discussed in the SES literature and also one of the original design principles—*collective-choice arrangements*—in "Governing the Commons"

(Ostrom 1990). Many authors see a clearly positive relationship between participation and success of the system. However, the exact nature of that relation remains unclear.

Opportunities for the participation of the involved actors is a prerequisite for the process of crafting rules for SES (Gutiérrez, Hilborn, and Defeo 2011). For example, the costs for rule changes decrease when the majority of those affected created the rule-system themselves, thus being in an optimal position to change or enforce it, too. In addition, rule compliance (F19) is higher when rules have been crafted by the actors' group itself and not imposed from outside (Kosfeld, Okada, and Riedl 2009; Sutter, Haigner, and Kocher 2008). With high participation, rules can be adapted quickly and targeted precisely—the congruence of requirements and compliance will increase as well (Ostrom 1990; Pomeroy, Katon, and Harkes 1998). Rules crafted by collective-choice arrangements ensure that benefits accrue to the actors themselves. Prior experience with collective-choice mechanisms enhances chances that the rules crafted are well designed for the purpose they fulfill (F9).

Rules are usually much clearer as well with participative action (F17, see Pagdee, Kim, and Daugherty 2006). This applies to all levels, for example, to gathering of information or to decisions or evaluations (Gruber 2008). If actors elect a local body to act in their place and their decisions are transparent, the effect is equally beneficial (Wade 1992). An operationalization of opportunities for participation can be found in (Shiferaw, Kebede, and Reddy 2008).

With participation processes organized at different levels each of them is targeted toward its purpose. The right choice of level ensures that problems can be solved efficiently, quickly, and accurately. Cross-linking these problem-solving levels ensures that the various rule systems do not remain isolated. There is a correlation between tenure regime and participation (Nagendra 2007), and a high dependency on resource system (F10) increases the likelihood of participation (Agrawal and Chhatre 2006).

Participation of the majority involved ensures fair and open processes. The involvement of actors can be encouraged through specific incentives (tax credits, assistance, training opportunities, etc., see Schurr 2006). On the other hand, structures in place prior to potential community management may sometimes constitute almost insurmountable obstacles (Schurr 2006).

A long-term plan for the future available and comprehensible to all, coupled with action arenas, improves chances of success. It has been suggested that consulting different experts on their respective area of expertise is of high relevance to success (Scheberle 2000).

A very real danger in many systems is that key decisions are made by a few rich or powerful individuals despite existing participation. Therefore, it is especially crucial for foreign aid projects that participation does not remain an empty phrase or is grafted to the system after implementation. It should be tied to the efforts in a central position to improve the system (Baland and Platteau 1996).

Connections to other concepts: social capital (F9), adapted rules (F17), fairness (F18)

F14—Legal certainty and legitimacy

Definition/Description: Acceptance of the local community as authority in regard to local jurisdiction and local legislative authority. The level of recognition of these rights by other groups, NGOs, and the national state ensures the stability and security of local transactions.

Reasons, why important for success: Legal certainty is a key prerequisite for long-term planning and thus sustainability (Ostrom 1990). In particular, property rights must be safe and stable (Schlager and Ostrom 1992). If such security is given, the willingness to invest in the resource increases considerably (F10). Legal certainty is the basis for long-term management policies which in turn are important for ecological success (Gutiérrez, Hilborn, and Defeo 2011).

On a national or international scale, legal certainty has positive effects on stability and, again, the willingness to invest in the resource system. It should be guaranteed that the state itself cannot override or ignore local rules. This is one of the original design principles in "Governing the Commons," namely *minimal recognition of rights to organize* (Ostrom 1990). This may include transfer of rights by national states to local communities (Pomeroy, Katon, and Harkes 1998). In case studies where this concept is not in place, like, for example, in Honduras or Guatemala, community-based tenure regimes cannot come into existence (Tucker, Randolph, and Castellanos 2007). Only within a framework of legal certainty, does collective action have the freedom and independence it needs to develop (Olsson, Folke, and Berkes 2004).

These rights can then be locally adapted (F17) with the certitude that the local legal system is not disturbed from outside (Wade 1992). The opposite situation leads to unstable systems and collapse. This has been demonstrated by the tacit agreement of the national government of Turkey allowing international trawlers to empty local fishing grounds (Berkes 1992).

In other cases, state regulations *replace* existing local rules—sometimes with the best intentions, sometimes not (Ostrom 1992a), but often destroying existing locally adapted rule systems (F17).

In general, many authors see a clear relationship of this factor to success:

> When security of tenure is absent—in other words, when user rights and benefits are insecure—CFM [community forest management is more likely to fail. (Pagdee, Kim, and Daugherty 2006, p. 44)

Legal certainty is most prominently visible in the recognition of property rights (Tucker 2010). Underlining this argument, Tang (1989) points out the crucial role of legitimacy to make penalties authoritative and binding. Without a legal basis, sanctions are simply not followed (F20). In fact, a small bundle of quite minor legal alleviations will usually suffice to increase incentives for collective action dramatically (Schurr 2006).

Connections to other concepts: relations (F23), compliance (F20), adapted rules (F17)

F15—Administration

Definition/Description: The administration communicates and executes rules and decisions of the rule-giving body. It is also responsible for parts of the organization of the community.

Reasons, why important for success: Implementing rules that are agreed upon and organizing the community is the job of an administration, which is thus central to many processes (Pomeroy, Katon, and Harkes 1998). A local administration has to be legitimized, enjoy the confidence of the actors, and assign clear responsibilities. If this is the case, compliance will benefit (F20). To be effective, individuals within an administration must make fair decisions and must not be corrupt. Ideally, employees take part in local life and know appropriation problems from their own experience (Ostrom 2009). Efficient management lowers the costs of organization (Tang 1989). A well-functioning administration ensures that organizational decisions are indeed implemented (Gruber 2008). Thus, an administration helps to translate institutional decisions into efficient and actual facts. However, efficiency often ultimately hinges on the skill and experience of a few administrators (Pagdee, Kim, and Daugherty 2006).

Connections to other concepts: compliance (F20), relations (F23), information (F16)

F16—Information

Definition/Description: The information and communication flow within the community and to other stakeholders. This includes actors' knowledge of the SES.

Reasons, why important for success: Information is required for assessing the condition of the resource system and regulating the flow of resource units. Only after information is acquired, it becomes possible to adapt rules and to control monitoring and appropriation (Sandström and Widmark 2007; Ostrom 2009). Good information is particularly important in systems with a high mobility of resource units, a heterogeneous distribution, and unpredictable system dynamics (F5) as well as a large size (F1). A strong information system generates knowledge about all parts of the system—resource, resource units, rule system, the actors' group itself, and other parties involved. It may also strengthen the cohesion of the social network (F9, see Gruber 2008). For that, it is crucial that the actors involved have easy access to existing information—such information has to be made public in an appropriate way (Schurr 2006). Without such knowledge, collective action will not get started at all—because neither costs nor benefits are calculable (Scheberle 2000; Pomeroy, Katon, and Harkes 1998; Thomson, Feeny, and Oakerson 1992). Particularly local knowledge (F17) should be included in such an information base and not be dismissed as unimportant. Unfortunately, local knowledge is still treated poorly in many foreign aid projects (Ostrom 1992a). Not surprisingly, dissemination of information correlates positively with economic and ecological success, for example, in Indian community watershed management (Shiferaw, Kebede, and Reddy 2008).

Part of a good information policy is open communication with all parties involved (F23). Ideally, arenas exist where free and unimpeded communication is possible in order to communicate different goals and values to other participants or groups (Pomeroy, Katon, and Harkes 1998). SES-management has even been described as an information network and classified as very information-intensive (Olsson, Folke, and Berkes 2004). Within this network, it must be distinguished between horizontal (e.g., between communities) and vertical links (e.g., from a community to the regional government). Finally, the amazing level of detailed system knowledge that Japanese communities developed for ecological relationships has been demonstrated (McKean 1992). Such knowledge may flow directly into very highly customized rules (F17).

Connections to other concepts: relations (F23), manageability (F5), resource size (F1)

F17—Characteristics of rules

Definition/Description: Rules are adapted to local circumstances. They are easy to understand, flexible to adapt to new situations, and there are feedback mechanisms about their effective operation.

Reasons, why important for success: Although systems may appear quite similar in geographical terms, type of resource, and appropriation processes, they nevertheless require different management strategies (Ostrom 1990). This is one of the original design principles—*congruence between appropriation and provision rules and local conditions*—of "Governing the Commons" (Ostrom 1990). In the past, many attempts to transfer a successful solution to similar systems have been unsuccessful (Meinzen-Dick 2007; Ostrom, Janssen, and Anderies 2007). This has been dubbed the panacea problem or blueprint fallacy.

It is therefore important to adapt control systems to local conditions (Baland and Platteau 1996; Pagdee, Kim, and Daugherty 2006). This can be demonstrated particularly well by comparing case studies (Tucker, Randolph, and Castellanos 2007). Cultural and social environmental parameters have to be taken into account; failure is almost inevitable if local customs are ignored or contradictory actions are taken (Pomeroy, Katon, and Harkes 1998). Ideally, rules integrate seamlessly into cultural and ideological givens and traditions (F9, see Cox 2010). Adaptability also includes specific biophysical circumstances and requirements. Flexible rule systems are able to respond to dynamic changes in both biophysical and social circumstances. Seasonal changes are a very simple dynamic change to be handled by different rule sets. There are examples of rotation systems addressing particularities of different spots (Berkes 1986).

Flexibly adapted rule sets have to be easy to understand, clear, transparent, and simple (Berkes 1992). Very complex provisions and laws may prevent collective action (Scheberle 2000). A major prerequisite for collective action is trust (F9) and good information management (F16) in particular concerning the condition of the resource and the actors' group (McKean 1992). Customized rules significantly increase efficiency and equity. Case studies demonstrate that sophisticated systems of monitoring and penalties (F19, F20) are sometimes in place for several hundred years (Casari 2003).

It has often been concluded that national governments are not suited to collect local information. For this reason, it has also been suggested that organizing the resource should be moved to the location in question. However, it is also conceivable to combine the strengths of a central instance (commanding resources like money, expertise, etc.) with the strengths of local institutions. This requires *nested enterprises*, one of the original design principles (Bloom and German 2000) to address problems on the right level. Redundancy may be positive in such cases (Gutiérrez, Hilborn, and Defeo 2011). One example could be states adopting laws facilitating collective action. Communities coming into existence because of this in turn create rules tailored to the special requirements of CPR systems (Schurr 2006).

Connections to other concepts: social capital (F9), monitoring (F19), exclusion (F22)

F18—Fairness

Definition/Description: This concept describes actions, behavior, and rules that should be just as well as level inequalities as far as possible. It comprises equality of benefits as well as equality of costs and duties.

Reasons, why important for success: It is well documented that fairness is a basic condition for the motivation of people to constructively participate in community actions (e.g., Falk, Fehr, and Fischbacher 2003). Unfair distribution destroys altruism quickly and completely (Fehr and Fischbacher 2003). If institutions are unfair, they are very likely to be rejected in the start-up phase (Kosfeld, Okada, and Riedl 2009).

Theoretically, it is difficult to define what this concept means (Rawls 1979), but in practice that has been solved: fairness is to be interpreted from the perspective of those affected (Fehr and Schmidt 1999). A distribution of units *proportional* to the efforts of the parties involved may be regarded as fair in one community. In another community this could be interpreted as unfair, preferring a distribution in which all get the *same* amount instead. Perspectives of what is fair may even differ within a community: the wealthy may often be in favor of proportional allocation while the poor prefer equal allocation (Dayton-Johnson 2000).

Fairness emerges in rules crafted by those in authority (Baland and Platteau 1996) but is also reflected in voting rights (Scheberle 2000). Moreover, inequalities can be compensated with appropriate, skillfully selected rules. For example, in Japan, harvested grass of differing quality is divided into clusters and distributed by lot (McKean 1992). Such a system is also used in Switzerland for logging piles. In fisheries, as well as in irrigation systems, where some spots are known to be better than others, rotating systems are common to ensure fairness (e.g., Berkes 1987). Additionally, precise accounting of deposit payments and withdrawals helps keeping the distribution of profits as fair as possible (McKean 1992).

A fair distribution of responsibilities and rights results in a higher willingness to invest in the resource (F10, see McKean 1992). If only few individuals benefit from the profits, reduced motivation and willingness to invest are the result (Thomson, Feeny, and Oakerson 1992). Interactions of the actors' group with the outside should be characterized by fairness as well; in particular it is important that they receive some form of compensation for the protection of the resource by the national government if they do so (Gruber 2008).

Connections to other concepts: social capital (F9), monitoring (F19), dependency on resource (F10)

F19—Monitoring

Definition/Description: Monitoring means gathering information about social and environmental aspects of appropriation. This information can then be used both for control and ensuring rule compliance.

Reasons, why important for success: If people know that they are or may be observed, they conform more to rules, even if cues are minimal (Bateson, Nettle, and Roberts 2006). Monitoring in SES is typically either performed by the actors themselves or by appointed guardians accountable to the actors (Gruber 2008). In agency-managed SES, the lack of guardians' motivation is often a problem. Such problems may be resolved in local systems by paying the guards with additional incentives to receive the penalties of those violating the rules (Wade 1992). Furthermore, the guards' payment should be high enough to discourage bribery— although de facto this is almost never the case (Baland and Platteau 1996).

Since SES are by definition always at risk of free riders, monitoring seems to be necessary in general. *Monitoring* is one of the original design principles in "Governing the Commons" (Ostrom 1990). If actors succeed in effectively deterring free riders through checks and penalties (F20), the effectiveness of the system can be greatly increased (Gutiérrez, Hilborn, and Defeo 2011).

Successful monitoring means providing adequate means as well as plans to deploy these means sensibly (Pomeroy, Katon, and Harkes 1998). The costs of monitoring are usually relatively small if the appropriators themselves monitor (Ostrom 1990). If this is the case, additional paid guards are not necessary— monitoring is done besides the actors' daily activities with almost no extra effort (Tucker, Randolph, and Castellanos 2007).

If the "shadow of the future" (Axelrod 1984/2000) is long enough, monitoring costs will decrease, for two reasons: first, participants start out with less punishment; second, the punishment level is consistently lower throughout. Therefore, participants take into account the duration of a cooperative interaction (Gächter, Renner, and Sefton 2008; Casari 2003), which may be considerably long in SES-management. Costs can be reduced to almost zero through appropriate arrangements. Rotating systems are but one example: each participant monitors his predecessor (Cox 2010). In some SES (here: forests), the length of monitoring is the most important positive factor (Agrawal and Yadama 1997). In contrast, another study by the same author finds a *negative* correlation with forest condition (Agrawal and Chhatre 2006). In any case, monitoring is a necessary precondition for rule compliance (F20)—it is impossible to detect violations without monitoring first (Pagdee, Kim, and Daugherty 2006).

The design of monitoring has to be "monitored" itself. Ideally, this is done by democratically legitimated bodies (McKean 1992). This conclusion is supported by a study able to demonstrate a relationship between the tenure regime of the community and monitoring, which in turn directly influences forest condition (Nagendra 2007). Summing up many other studies, Tucker (2010) concludes that monitoring is one of the key factors for ecological success, at least for forests.

Connections to other concepts: compliance (F20), conflict management (F21), information (F16)

F20—Compliance

Definition/Description: The level of complying with established rules and legislation. In many systems, compliance is achieved through enforcement (e.g., guards) and sanctions.

Reasons, why important for success: Even in properly functioning rule systems, violations do occur. Humans are quick to identify gaps and calculate costs and benefits of rule violations if monitoring is sporadic or absent. It is therefore essential for compliance that penalties are *actually* and *effectively* enforced and do not exist only on paper (Pagdee, Kim, and Daugherty 2006; Ostrom 1990; Gibson, Williams, and Ostrom 2005). Moreover, if imposed penalties are only rarely put into effect, compliance will decrease rapidly. This may lead to the exploitation of the resource (cf. Vollan 2008). Compliance strongly depends on whether rules and enforcing institutions are perceived as legitimate (Baland and Platteau 1996).

In many SES, *graduated sanctions* (one of the original design principles in "Governing the Commons," Ostrom 1990) have a proven track record: depending on the nature, severity, and repetition of the offence, relatively mild penalties are imposed at first but harder in due course. Why do even mild penalties work so well? They often mean a loss of reputation for the perpetrators and a reputation gain for the reporter (Wade 1992; Ostrom 1990). Many studies show that the possibility of building up a reputation boosts cooperation (Price 2006; Wedekind and Milinski 2000; Rockenbach and Milinski 2006; Resnick et al. 2006).

Punishment strengthens the rule system and ensures the cohesion of the community, which in turn is relevant for stability (F9). Sanctions also prevent abuse, which adversely affects the condition of the resource (Cox, Arnold, and Villamayor Tomas 2010). However, some authors argue that the presence of penalties could well be an indicator for failure since sanctions should not be necessary if there is strong leadership (Lam 1998; Cleaver 2000). It is best if enforcement is performed by the actors' group itself (Gruber 2008; Gutiérrez, Hilborn, and Defeo 2011; Ostrom 1990), because actors are usually vigilant and effective, since they directly benefit from a well-maintained resource (Pomeroy, Katon, and Harkes 1998).

Another study suggests a relationship between clear rules (F17) and compliance (McKean 1992): only clear rules allow an accurate assessment of whether appropriation is near the maximum sustainable yield, that is, whether it is efficient. If this is the case, rule compliance is higher.

Rule enforcement is linked to resource size (F1)—very small communities may not need material sanctions at all, relying instead only on reputation losses and gains (Agrawal and Chhatre 2006). A study of nine forests in Central America sees rule enforcement as the *sine qua non* (Tucker, Randolph, and

Castellanos 2007). The positive correlation between compliance and condition remains statistically reliable even with larger samples in forests worldwide, which demonstrates the importance of this concept (Tucker 2010; Gibson, Williams, and Ostrom 2005). Other studies demonstrate that enforcement has to be locally adapted (F17), too, because its effectiveness depends on that very fact (Casari 2003). However, Chhatre and Agrawal (2008) suggest that although this factor is of indisputable importance, it is mediated and modified by a multitude of others.

Connections to other concepts: monitoring (F19), conflict management (F21), resource size (F1)

F21—Conflict management

Definition/Description: How disagreements between individuals, groups or other stakeholders are treated, solved or mediated both within the community and with outside parties.

Reasons, why important for success: Even clear and simple rules can be interpreted differently and are often subject to dispute. Hence, these disputes have to be settled by a third party, for example, a local court or any other arena. A solution of such conflicts contributes to the stability of rule systems and thus a stable use of the resource (Ostrom 1990). It is also one of the original design principles—*conflict-resolution mechanisms*—in "Governing the Commons" (Ostrom 1990). It is important that conflicts are resolved quickly at low cost and a jurisdiction which is locally available and easily accessible for everyone. An example would be arbitration boards (Ostrom 1990). Such jurisdiction should work fairly and without corruption. If conflicts do not escalate, cooperation and trust are strengthened, coordination is furthered, and efficiency is increased. Typical causes for conflicts are a rapidly growing population (F7), which increases the pressure on the resource (F6), a high heterogeneity of actors (F8, see Baland and Platteau 1996), or new migrants demanding the same rights as long-term residents.

There seems to be a link between the tenure regime (leasehold) and a higher incidence of forest management conflicts (Nagendra 2007). Violence is—as expected—correlated with poorer forest condition, while a battle of the sexes show—at least in one study—a positive effect (Agrawal and Chhatre 2006). The suggested cause may be that women get participation rights only after heavy conflicts while participation as such is ultimately positive for forest condition. However, other authors do not find a significant correlation between conflicts and forest condition (Pagdee, Kim, and Daugherty 2006). In general, local conflict resolution involving all stakeholders seems to be more successful than conflict resolution by the state (Pomeroy, Katon, and Harkes 1998). It is also advantageous, of course, to have contracts and conflict mechanisms already in place. They should be agreed on by the majority before any serious disputes happen within the community.

Connections to other concepts: compliance (F20), number of actors (F7), group composition (F8)

5.1.5.5 External influences
F22—Exclusion of third parties

Definition/Description: Exclusion of third parties refers to preventing appropriation by other groups through law or rules.

Reasons, why important for success: The importance of this factor follows directly from the definition of CPR, a frequent management form of SES: if there is no exclusion or if it is not viable, it is no longer CPR management, but open access. This is a fundamental distinction Hardin (1968) failed to make (Ostrom 2009) which led him to his famous but deeply pessimistic conclusions which many scholars have equaled with concluding that CPR are necessarily doomed to fail as well. Indeed, open access regimes lead almost always to the destruction of the resource, as no actor has a reason to preserve it, since preservation is advantageous only to others. The exclusion of third parties is therefore central to stability and environmental sustainability. In fact, every CPR analysis has to start with this question (Berkes 2007).

The transition from open access to CPR management may be driven by the *possibility* of excluding third parties with technical measures such as fences or rule-bound measures such as property rights. For example, in some Indian villages only sheep of actors are allowed to graze on the local community pastures, but no sheep of outsiders (Wade 1992). In the case of exclusion by technical means it is important that these means are cheap or easy to implement and that the act of exclusion does not lead to conflicts. If the exclusion of third parties is given the likelihood increases that actors will manage the resource carefully and, by themselves, develop rules for its protection.

This concept is therefore particularly important for concrete appropriation decisions (Feeny 1992). In the case of small Turkish fisheries, the problem of not being able to exclude third parties (large fishing trawler who completely exhaust fish stocks in one haul) is even dangerous to their very existence and lead to the failure of local management (Berkes 1992). There are different ways of regulating that only a certain group is entitled to appropriate (Tang 1992). Such *boundary rules* include land ownership, membership of an organization, special permissions, or payment of a fee for appropriation. Since the superiority of any of these mechanisms has not yet been proven, it does not seem to be decisive which of these mechanisms is used. It is more important whether such rules are considered fair and legitimate and if they are locally adapted (Ostrom 1990).

Connections to other concepts: resource boundaries (F2), group boundaries (F12), legal certainty (F14)

F23—Relations with external parties

Definition/Description: This concept describes the various interactions between all involved parties like appropriators, NGOs, the national state, and other nearby appropriating groups. These interactions span a continuum between close collaboration (partnership) and heavy conflict.

Reasons, why important for success: Cooperation with either the state or organizations such as NGOs strengthens legitimacy and legal certainty (F14), which in turn promote long-term investment in the system (F10, see Berkes 1992, 2007). Since each stakeholder group is located within a network of relationships with other parties the manner of these relations is, according to some authors, absolutely critical for success or failure (Pomeroy, Katon, and Harkes 1998). This claim is supported by the fact that good partnerships can lead to innovations and a more productive use of resources (Gruber 2008).

Large projects are often only feasible in cooperation with the state. Then, a good relationship with the respective national state becomes vitally important. Good state relations are also relevant for loans on special terms by the state (Baland and Platteau 1996). If other parties aid with concrete technical assistance, efficiency is usually increased greatly because problem situations can be identified and addressed accurately by experts who are familiar with modern solutions for typical problems (Pagdee, Kim, and Daugherty 2006). Financial aid improves the cost-benefit balance of appropriators. One important kind of aid is the compensation for nature protection measures by the state, or the designation of protected areas in the vicinity, which usually means reduced earnings for the actors (Gutiérrez, Hilborn, and Defeo 2011). However, two aspects of outside help often become problematic: first, if no attention is paid to the existing local institutions (F17) and second, if the focus is exclusively on improving technology without regard to institutions. Unfortunately, many foreign aid projects still commit these errors (Ostrom 1992a).

To conclude: the diverse interactions with various other organizations or groups are highly relevant for success (Ostrom 2009; Schurr 2006). However, Agrawal (2001) quite rightly points out that external relations in CPR research are neglected in many studies. In his opinion, this is due to historical reasons: since CPR research had to establish first that collective action is important, a fact that has been widely ignored by traditional economic research. Therefore, the focus has been on the respective local community and not on other stakeholders.

Connections to other concepts: legal certainty (F14), participation (F13), administration (F15)

F24—Capability to adapt to changes (resilience)

Definition/Description: System and actors are able to cope with sudden changes in technology, in interaction with markets or in the natural environment.

Reasons, why important for success: Sudden changes—for example, in technology, in interaction with markets, or natural disasters—can disrupt the delicate balance in SES (Xu, Marinova, and Guo 2015). Rule systems have to be flexible to be able to deal with sudden shifts or crossing certain social or ecological thresholds (Blythe 2015). If they are able to do this, they are *resilient* (Folke 2006; Anderies, Janssen, and Ostrom 2004). Such adaptive systems have, in comparison, a better long-term perspective because they are more stable in crises, thus increasing their chances of survival.

With regard to technological innovations, sudden changes in appropriation technologies seem to be detrimental. Technologies should not change too abruptly but be consistently used over longer periods of time. It helps, too, if a transitional period for the introduction of new technologies enables a slow, gradual transition. Whenever new technologies are implemented, particularly costs and benefits of extraction are affected. In the worst case this may mean that the entire system has to be rebuilt (Agrawal 2002).

Market integration plays a role as well. Are markets for the appropriated units easily accessible? Are there strong price fluctuations for the product? Are markets linked to others? If no market is available or if it is difficult to access, the extracted product has to be consumed or processed by the actors themselves, which may not be optimal in terms of efficiency or benefits. Low market availability increases dependency on resource (F10). However, selling the extracted units on markets may be linked to high dependency, too. Whenever strong price fluctuations in dynamic markets increase the uncertainty of harvests, stability is adversely affected, because overharvesting is incentivized by high prices, which in turn enhance further disregard for sustainability considerations. In sum, a weak connection to stable markets may therefore be positive.

Since many aspects of the environment tend to change more rapidly in a globalized world, adaptability plays an ever increasingly important role (Baland and Platteau 1996). It has also been pointed out that the complex interplay of many factors contributes to the change of institutions in such a rapidly changing environment (Pagdee, Kim, and Daugherty 2006). It is mostly the leaders's responsibility to adjust local rules to new circumstances. Interestingly, local communities tend to adopt new technologies reducing the pressure on the resource more readily than other forms of management (Nagendra 2007). This concept (like F23) tends to be underrepresented in SES-research as the focus of most studies is on the local system itself and less on the environment (Agrawal 2001).

Connections to other concepts: condition at the beginning (F4), regeneration (F6), information (F16)

5.2 Results for the Common-pool resource data

Anticipating the most important results, the CPR models show very high explanatory value, especially the neural networks. Furthermore, all three methods mostly are in agreement about the importance of factors, assigning legal certainty (F14) the most important role.

The presentation of the empirical model results—the core of this book—is arranged by data set, so that one section is dedicated each to CPR, NIIS, IFRI, and the overall model (see Sections 5.2, 5.3, 5.4, and 5.5), starting with the CPR. The respective discussion is directly connected to these results. Each section is structured identically: First, the data set is characterized by descriptive statistical information (see also Table 7.5 in the Appendix) and correlations (see also Table 7.6 in the Appendix). This provides an initial overview. This is followed by the results of each modeling method—linear regressions, random forests, and neural networks. Another part is taken up by robustness and sensitivity analyses (Section 5.6). The comparison of modeling results is then presented in more detail in the discussion (see Chapter 6).

Besides model performance, each section discusses the relevance of the individual success factors for each of the three methods used. For regressions, these are the standardized beta coefficients, for random forests this is the explicit ranking by factor importance (see Section 4.1.2), and for neural networks these are the extracted weights (see Section 4.1.3.4).

These result sections are supplemented by an overview of all other model variants (Section 5.6). Due to the high number of calculated models, a robust overall estimation of the importance of each factor is possible. For these variants, the mean importance for each factor is calculated for each statistical method (MLR, RF, NN) and each data set (CPR, CPR Top 3, NIIS, NIIS Top 3, IFRI, IFRI Top 3).

5.2.1 Descriptive statistics

All descriptive statistics for all data sets are available in the Appendix—statistical key figures as well as histograms (Figures 7.1–7.4) visualizing the distribution across cases for each success factor.

The success factors for the 122 CPR cases are distributed relatively evenly (Appendix, Section 7.5.2.1), but not as evenly as IFRI or NIIS. Exceptions are some factors with low minima: −0.3 for resource limits (F2); 0.08 for group composition (F8); −0.23 for legal certainty (F14); and -0.25 for relations to others (F23).

The mean and median for the factors accessibility (F3), number of actors (F7), group composition (F8), administration (F15), and relations to others (F23) is shifted in a positive direction. There are two reasons for this: one, systems in this data set have not been selected randomly with regard to these factors, that is, there are significantly smaller and more homogeneous groups than in the other data sets; two, for some factors only a few variables are available, which may shift their measures of central tendency. Only the factors number of actors (F7) and group boundaries (F12) show a high kurtosis, all other factors are distributed normally, which is confirmed by the variance, which show small deviations except for fairness.

The dependent variable, ecological success, has a mean value of -0.17 and is therefore almost exactly 0, even if the maximum is very low at only 0.26. The selected systems are thus close together with a variance of only 0.06; the worst system performance is -0.77. Compared to IFRI and NIIS, the average success is thus the lowest of all data sets.

5.2.2   Correlations

Further information on the data structure is provided by the correlations of success factors. Figure 5.2 (and Section 7.5.2.3 in the Appendix) describe the Pearson correlations of CPR data set factors. The numbers in the tables represent the correlation coefficients, the asterisks mark the significance level (* = p<0.05; ** = p<0.01; *** p<0.001). Note that positive correlations are marked clockwise beginning from the top, while negative correlations are marked in a counterclockwise way in all correlational matrices.

The exact values of all correlation matrices can be found in the Appendix, Sections 7.5.2.3, 7.5.3.3, 7.5.4.3, and 7.5.5.3.

The CPR data set shows 132 significant correlations between the 24 factors themselves and ecological success (Appendix, Section 7.5.2.3). The strongest correlations are between governance systems and actors, which also corresponds to theoretical expectations. All correlations are Pearson correlations, significant at the p<0.001 level if not stated otherwise. Note that some correlations are described as if there is a causal relationship ("leads to," "contributes to"). These are based on the literature review in Section 5.1.5, and are not deduced from the correlations.

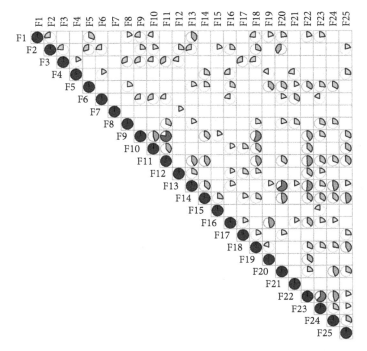

Figure 5.2 Pearson correlation coefficients between success factors of CPR

Note: Positive correlations are shown clockwise, negative correlations counterclockwise.

Thus, fair rules (F18) are accompanied by high social capital (F9; Pearson r = 0.55; all further correlation coefficients are Pearson coefficients), just as legal certainty (F14) entails high compliance (F20; r = 0.47). It is an important precondition for the formation of social capital (F9) if the individual actors depend on their group (F11). There is also a high correlation between social capital (F9) and dependency on group (F11; r = 0.78). This dependency on group usually also means high participation possibilities (F13) and the ability to exclude other actors (F22; r = 0.42 and r = 0.49). The literature discussion also confirms (see Section 5.1.5.4) that high participation possibilities (F13) lead to high compliance (F20; r = 0.65) and that secure basic rights (F14) also contribute to compliance (r = 0.36).

These results confirm predictions and results from the literature and show the quality of the data. The fact that all correlations are positive is a strong indication that the direction of factors was chosen correctly. For example: less actors (F7) is coded with 1, more actors with –1.

Apart from uncovering important relationships *between* success factors, correlational analysis can also show direct connections to success. In the CPR data set, thirteen factors are directly and significantly correlated with success:

- clear resource boundaries (F2; $r = 0.20$, $p<0.05$),
- initial ecological success (F4; $r = 0.26$, $p<0.01$),
- social capital (F9; $r = 0.32$),
- dependency on resource (F10; $r = 0.34$),
- dependency on group (F11; $r = 0.39$),
- participation (F13; $r = 0.20$, $p<0.05$),
- legal certainty (F14; $r = 0.48$),
- locally adapted rules (F17; $r = 0.25$, $p<0.01$),
- fairness (F18; $r = 0.42$),
- compliance (F20; $r = 0.31$),
- exclusion (F22; $r = 0.20$, $p<0.05$),
- relationships with others (F23; $r = 0.18$, $p<0.05$) and
- capability to adapt to change (F24; $r = 0.31$).

Some of these, namely clear boundaries (Wade 1994), dependency on resource (Agrawal 2007), legal certainty (Schlager and Ostrom 1992), fairness (McKean 1992), opportunities for participation, social capital, locally adapted rules (Ostrom 1990), and compliance (Pagdee, Kim, and Daugherty 2006) have already been explicitly associated with success. After this explorative data analysis, I turn to regressions which are the first of the three statistical methods used to create a model.

5.2.3 Multivariate linear regressions

The regression shown in Table 5.2 refers to the fully weighted and imputed CPR data set with all variables. The results of all other model variations are shown in Section 5.6.1.

The model summary in Table 5.3 shows model quality (adjusted R^2) and other key numbers. Note that all R^2 reported in the text are always adjusted R^2.

Besides the full model, the most *parsimonious* model with sufficiently high explanatory power is also of interest. Two methods were used for all data sets (CPR, NIIS, IFRI, combined model)—backward elimination (removing variables with smallest partial correlation to the independent variable) and automatic linear modeling. Summarizing the results, the model with the fewest factors still has fourteen factors (F2, F6, F8, F9, F10, F11, F12, F14, F17, F18, F21, F22, F23, and F24) and an adjusted R^2 of 0.497. On the other hand, the most parsimonious model with only five dependent variables has an adjusted R2 of 0.39. Fairness (F18), legal certainty (F14), clear boundaries (F2), dependency on resource (F10), and the regenerative capability of the resource units (F6) are important here.

Table 5.2 Multivariate linear regression for the complete CPR data set

Model MLR	Non-standardized coefficients		Standardized coefficients	T	Sig.	95.0% Confidence-interval for B		Collinearity statistics	
	Regression coefficient B	Standard error	Beta			Lower limit	Upper limit	Tolerance	VIF
(Constant)	-.273	.096		-2.837	.006	-.464	-.082		
F1	-.005	.032	-.016	-.172	.864	-.068	.057	.480	2.085
F2	.373	.097	.375	3.845	.000	.180	.565	.442	2.260
F3	.048	.050	.080	.959	.340	-.052	.148	.608	1.646
F4	.009	.043	.018	.222	.825	-.075	.094	.624	1.603
F5	-.066	.063	-.101	-1.036	.303	-.192	.060	.439	2.279
F6	.168	.042	.370	3.973	.000	.084	.252	.485	2.061
F7	.062	.039	.122	1.593	.114	-.015	.140	.722	1.385
F8	-.186	.084	-.188	-2.223	.029	-.352	-.020	.588	1.701
F9	-.074	.047	-.206	-1.563	.121	-.168	.020	.241	4.143
F10	.188	.059	.270	3.189	.002	.071	.305	.584	1.712
F11	.153	.056	.382	2.750	.007	.043	.264	.217	4.603
F12	-.200	.067	-.270	-2.984	.004	-.333	-.067	.512	1.954
F13	-.002	.075	-.004	-.032	.974	-.150	.146	.249	4.014

Continued

Table 5.2 *Continued*

Model MLR	Non-standardized coefficients		Standardized coefficients	T	Sig.	95.0% Confidence-interval for B		Collinearity statistics	
	Regression coefficient B	Standard error	Beta			Lower limit	Upper limit	Tolerance	VIF
F14	.218	.128	.170	1.697	.093	-.037	.472	.418	2.395
F15	-.005	.035	-.012	-.149	.882	-.075	.065	.601	1.664
F16	-.033	.044	-.072	-.745	.458	-.120	.055	.450	2.225
F17	.141	.054	.214	2.612	.010	.034	.248	.624	1.602
F18	.053	.041	.130	1.299	.197	-.028	.133	.420	2.380
F19	-.021	.067	-.030	-.316	.753	-.155	.112	.480	2.083
F20	.077	.058	.149	1.328	.187	-.038	.193	.333	3.006
F21	-.063	.035	-.151	-1.778	.079	-.133	.007	.582	1.718
F22	-.106	.080	-.162	-1.333	.186	-.264	.052	.284	3.520
F23	.264	.109	.260	2.433	.017	.049	.480	.368	2.719
F24	.096	.045	.201	2.138	.035	.007	.185	.475	2.103

Table 5.3 Explanatory power of the multivariate linear regression for the complete CPR data set

R	R^2	Adjusted R^2	Standard error of estimator	Change in R^2	Change in F	df1	df2	Sig. change in F
.770	.592	.491	.162	.592	5.873	24	97	.000

The next step is to determine the individual relevance of success factors to sustainability. Contributions of individual factors can be estimated using the standardized beta coefficients (Table 5.2). Eight factors have an above-average significance: resource limits (F2), regenerative capacity of the resource (F6), dependency on resource and group (F10 and F11), group limits (F12), local rules (F17), relationships with third parties (F23), and ability to adapt to change (F24). For all these factors, the beta coefficient is above 0.2; for group boundaries it is negative. Clear group boundaries that lead to the exclusion of others are therefore negatively linked to ecological success by MLR.

The importance of these eight factors is underlined by the fact that a regression model, in which only these factors are used as independent variables, performs only slightly worse than the model with twenty-four factors. Predictive quality (R^2) of 0.37 is only 0.12 below the overall model. Of these factors, legal certainty (F14) and fairness (F18) in particular are decisive in regression models. A regression with only these two factors still achieves an impressive R^2 of 0.31.

The test for multicollinearity of the factors is performed using the tolerance value or its reciprocal value VIF (variance inflation factor; last two columns). As a rule of thumb, a tolerance value of less than 0.1 or a VIF of more than 5 (or 10) is an indication of multicollinearity (Bortz and Schuster 2010, p. 350). Since no value is higher, it can be assumed that there is no multicollinearity. This also applies to NIIS, IFRI, and the overall model.

Since the predictive accuracy of statistical test procedures such as random forests and neural networks is usually not calculated using the entire data set but on one test data set, this procedure is also performed for the regressions for comparability. Table 5.4 summarizes model performance for different splits between training and test data set (see Section 3.5.5).

Model performance both on the training and test set data does not differ very much and confirms the robustness of MLR. The two exceptions are the split for social capital with 0.29 and for size with 0.69. Hence, predictive accuracy in the test set approximately corresponds to that in the training set.

Table 5.4 Explanatory power of the multivariate linear regression for the different splits of the CPR data set

Split (training and test set)	Adjusted R^2 on the training set	Pseudo-R2 on the test set
Relations	0.48	0.46
Ecological success	0.47	0.52
Fairness	0.47	0.49
Size	0.41	0.69
RU-Type	0.48	0.50
Social capital	0.53	0.29

Note: The parameter pseudo-R^2 tries to simulate model quality, which is expressed by the parameter R^2 (proportion of variance elucidation of the dependent variable by independent ones; cf., e.g., Williams 2011).

5.2.4 Random forests

The models of the random forests are also calculated using the different splits of the data set (Table 5.5, see also Section 3.5.5). The highest explanation of variance is 48%. The highest pseudo-R^2 on the test set is 77%.

Model performance of random forest models are slightly lower than those of neural networks, but are comparable to MLR (Table 5.5). Depending on split, performance ranges from 36% (size) to 45% (RU type). A model with all data explains 48% of the variance. The mean across all models is 42% on training data and 0.57 on test data sets. Therefore, model performance of random forest models is very similar, which speaks for the robustness of the method, similar to regressions.

Switching from overall model quality to the relevance of each influencing factor shows a very good agreement with the results from the MLR. First, however, the ranking criterion has to be explained in random forests. It is the percentage increase of the mean squared error (MSE) at random change of a variable. The higher the percentage increase in error (MSE), the more important the success factor for the model. This is a standard measure for random forests.

To make results more robust, the presented results for all random forest models on all data sets are an aggregation of the seven models discussed previously (six splits and the full data set, see Table 5.5) and models that have converged after six runs with different random seeds. The ranking from Table 5.6 is therefore an aggregation of all seven calculated random forests and thus a very robust measure of the importance in the CPR data set.

Table 5.5 Explanatory power and key data of random forests for CPR

Split (training and test set)	% explanation of variance	Mean square deviation	Number of trees	Pseudo-R^2 on test data	Number of variables used at split
Relations	37.89	0.033	1000	0.66	8
Ecological success	44.88	0.034	500	0.60	8
Fairness	45.05	0.033	500	0.51	8
Size	35.87	0.036	500	0.77	4
RU-Type	45.40	0.034	1000	0.44	16
Social capital	45.35	0.033	1000	0.42	8
Total data set	48.09	0.031	500	– (due to being the total data set)	8

The first column shows the success factors ordered descending according to importance (second column).

The most important factors for success across all random forests—all imputed (CPR unweighted, CPR Top 3 unweighted, CPR indicator weight, CPR Top 3, CPR fully weighted, CPR Top 3 fully weighted, CPR individual weighting fully weighted) are therefore fairness (F18), legal certainty (F14), good compliance (F20), and initial success (F4). According to these models, the first two factors are exceptionally relevant. Group boundaries (F12) and accessibility (F3) do not play any role. The ability of random forests to calculate the individual relevance of each success factor is particularly important because this can only be measured indirectly in neural networks.

Compared to the regressions, three of the five most important factors in the MLR also appear in the top group of random forests. Agreement with the two most important factors of the neural network models (see next section, Section 5.2.5), legal certainty (F14), and fairness (F18) is particularly evident.

5.2.5 Neural networks

This section summarizes model results for the neural networks. As in the previous section, models are presented for the different splits. Table 5.7 shows the best models for CPR.

Table 5.6 Importance of success factors for random forests for CPR

Success factor	% increase in MSE
F18 Fairness	20.87
F14 Legal certainty and legitimacy	17.20
F20 Compliance	15.70
F4 Ecological success at the beginning	13.39
F2 Resource boundaries	12.13
F9 Social capital	11.21
F11 Dependency on group	10.70
F24 Capabilities to adapt to change	10.49
F5 Manageability	9.23
F1 Resource size	9.12
F6 Regeneration of RU	9.07
F23 Relations	8.88
F8 Group composition	8.46
F17 Characteristics of rules	8.45
F13 Participation	7.51
F16 Information	7.27
F19 Control	7.09
F22 Exclusion	6.72
F15 Administration	6.52
F10 Dependency on resource	6.43
F21 Conflict management	6.02
F7 Number of actors	5.00
F12 Group boundaries	4.86
F3 Accessibility	4.48

Table 5.7 shows that splits by success or fairness achieve the best model quality for the test data (79% each). The smallest average absolute error (MAE) corresponds to these values.

In order to avoid that these results could also have resulted from non-reproducible lucky strikes, training runs are examined for *robust* patterns: only if a certain net configuration achieves the same model quality three times within

Table 5.7 Explanatory power and key data for the best neural networks for different CPR test data splits

	Relations	Success	Fairness	Size	Type of RU	Social capital
Number of neurons in hidden layer	29	18	26	18	16	27
Epochs	200	150	200	100	300	350
Algorithm	RPROP	RPROP	RPROP	RPROP	RPROP	RPROP
Min. error	−0.34	−0.33	−0.20	−0.26	−0.22	−0.28
Max. error	0.24	0.21	0.18	0.22	0.14	0.31
MSE	0.02	0.02	0.01	0.02	0.01	0.02
MAE	0.12	0.08	0.10	0.11	0.09	0.10
R^2	0.74	0.79	0.79	0.76	0.77	0.54

narrow tolerance limits (see methods in Section 4.1.3), the pattern found is considered stable and confirmed. Thus, all runs that have a certain network configuration (same learning algorithm, same number of neurons in its hidden layer, same split, and same number of epochs) perform very similarly regarding error rate and pseudo-R^2 if their training is repeated two more times.

These results are shown in Table 5.8, which shows the mean values of three runs per split.

Here, too, the splits according to success and fairness achieve the best predictive accuracy. Artificial neural networks are therefore significantly better in model quality than both MLR and random forests for the CPR data. While the first two methods explain about 40 to 50% of variance, the best networks achieve up to 79% (Table 5.7). Again, the six splits allow results to be classified in terms of their robustness: Although the split along social capital drops significantly (0.54) in pseudo-R^2, all other splits range between 0.74 and 0.79. Thus, these results point to an estimate what is achievable in model quality for the CPR data at all.

These results remain stable even for robust networks (Table 5.8). The best robust net is split according to success, has 21 neurons in the hidden layer, has been trained with the RPROP algorithm, has been trained for 250 epochs, and achieves a very remarkable R^2 of 0.71 (Table 5.8). Similarly, the best non-robust net has 26 neurons, has also been trained with RPROP with 200 repetitions, and reaches a pseudo-R^2 of 0.79.

It is difficult to extract the importance of individual factors from neural networks. The associated calculations are very computational-intensive and took

Table 5.8 Explanatory power and key data for robust neural networks for different CPR test data splits

	Relations	Success	Fairness	Size	Type of RU	Social capital
Number of neurons in hidden layer	29	21	13	7	28	26
Epochs	250	250	175	150	325	125
Algorithm	RPROP	RPROP	RPROP	RPROP	RPROP	RPROP
Min. error	−0.37	−0.41	−0.26	−0.3	−0.29	−0.26
Max. error	0.38	0.26	0.18	0.29	0.27	0.40
MSE	0.03	0.02	0.02	0.02	0.02	0.03
MAE	0.14	0.11	0.12	0.12	0.09	0.13
R^2	**0.60**	**0.71**	**0.68**	**0.70**	**0.68**	**0.38**

several days of pure computing time. The main disadvantage is the high level of uncertainty involved. For this reason, I combine the results (rankings of the importance of all factors) of all splits to an overall mean result including the mean standard deviation.

Using this method (see Section 4.1.3.4), the relevance of the factors in the models of neural networks can be determined. In Table 5.9, the factors are arranged in descending order of importance (column 2). As with random forests, the results of seven models are combined for the overall importance (all imputed: CPR unweighted, CPR Top 3 unweighted, CPR indicator weight, CPR Top 3 indicator weight, CPR fully weighted, CPR Top 3 fully weighted, CPR individual weight fully weighted). The uncertainty is expressed by the total mean value of the standard deviation.

The factors legal certainty (F14), resource boundaries (F2), and group composition (F8) are most important, followed by success at the beginning (F4) and the regeneration of the resource units (F6). Except for compliance (F20) and fairness (F18), all factors that MLR and random forests have calculated as important are also relevant here. What is striking is the absence of the factor fairness (F18), which plays only a very minor role in these models. As with the other data sets, further interpretation of these results is given in the overall conclusion in Chapter 5.

Table 5.9 Importance of success factors for neural networks for CPR

Success factor	Mean relevance (total)	Mean standard deviation (total)
F14 Legal certainty and legitimacy	15.38	6.87
F2 Resource boundaries	14.79	6.65
F8 Group composition	13.86	7.03
F4 Ecological success at the beginning	13.75	6.75
F6 Regeneration of RU	13.66	6.70
F12 Group boundaries	13.58	6.85
F23 Relations	13.52	7.09
F24 Capabilities to adapt to change	13.45	6.66
F13 Participation	13.37	6.76
F20 Compliance	13.03	6.79
F22 Exclusion	12.83	6.84
F11 Dependency on group	12.28	6.84
F10 Dependency on resource	12.15	6.71
F19 Control	12.09	6.77
F17 Characteristics of rules	12.02	6.67
F5 Manageability	11.91	6.93
F18 Fairness	11.59	6.60
F3 Accessibility	11.50	6.67
F1 Resource size	11.44	6.50
F16 Information	11.26	6.51
F9 Social capital	11.20	6.55
F21 Conflict management	10.73	6.48
F7 Number of actors	10.61	6.58
F15 Administration	10.02	6.49

5.2.6 Discussion

The following interpretation attempts to combine the many individual results into an overall picture. First of all, the CPR data set achieves the highest explanatory quality of all three data sets. This data set has very few successful systems, which is evident from the low maximum of 0.26 for success, even if the mean value is quite balanced at zero (−0.17).

There are close connections between social capital and fairness as well as between social capital and dependency on group. High participation opportunities go hand in hand with good compliance with rules, and good relations with other actors imply good opportunities for exclusion. Especially group characteristics and what they depend on (F10–F13) determine exclusion possibilities and correlate with success. These relationships show very close relationships between the composition of the group—dependency on it, common past and traditions, and clear group boundaries—and the rules that these communities make for themselves. Finally, there is a relatively large number of small systems which are able to set up a relatively conflict-free management with a homogeneous group, fair rules, and the successful exclusion of other users.

What is striking about this data set is the very good agreement of all three methods with regard to the significance of individual success factors for ecological success. Both multivariate regression and random forests consider two factors to be decisive: fairness (F18) and legal certainty (F14), while for neural networks it is only legal certainty.

It has been pointed out that especially for fisheries, legal certainty is crucial (Berkes 1986): local fishermen run the risk of losing their livelihoods through large trawlers. The latter are preferred by the state for reasons such as the promotion of national GDP. This is confirmed here.

Fairness, as the second important factor, plays a role for example for the fair allocation of fishing spots that may differ widely in profitability. For irrigation systems, making up the other half of case studies in the CPR data set, it has also been repeatedly pointed out that fairness is crucial. Fair rules have to compensate for the biophysical gap in water provision between head and end of irrigation systems (Tang 1992; Lam 1998).

Clarity of resource boundaries (F2) is also considered important for success in all three models. This directly refers to the character of shared resources. If the boundaries are unclear, people who do not belong to the group may profit. For example, it is difficult for fishermen to enforce their rights because it is questionable whether they even have these rights in a particular location. This relates directly to legal certainty. Only the combination of having clearly defined rights for a clearly defined area makes it possible to successfully defend against free riders. Fairness is also important for group-internal issues: If rules are unfair (F18), or if they are not followed for other reasons (F20), this usually means that fixed maximum harvest limits are exceeded, protected areas are violated, or other ecologically meaningful restrictions are not observed. This has a direct negative effect on ecological success. In combination with the correlations, the models see a high ecological success for rather small groups with high social capital and good opportunities for participation, operating in a secure legal area, and being strongly dependent on the group itself.

Overall, regressions and random forests explain about half of the variance, depending on split. Neural networks excel and even the robust networks achieve a goodness-of-fit of up to 0.71.

This means that very good models are available for the CPR data, which can explain a large part of the variance via success factors. The unexplained variance could, for example, come about through economic framework data, political, social and cultural influences, and peculiarities on a national level.

5.3 Results for the Nepal irrigation institution study data

For the NIIS data, the most important results are that all three methods agree on the three most important factors—participation (F13), dependency on resource (F10), and manageability (F5). Model quality is very good, explaining between 40% and 60% of the variance, although not quite as good as for the CPR. Analogously to the CPR, the NIIS data set is analyzed by presenting first the descriptive statistics, then the correlations, followed by the results of the three modeling methods.

5.3.1 Descriptive statistics

The descriptive statistics for the NIIS data set are consistently inconspicuous (see Appendix, Section 7.5.3 and Tables 7.7 and 7.8 for the corresponding histograms and statistical key figures). The success factors are more evenly distributed than in the CPR data set. The distribution of the factors F7, F9, F12, and F17 is skewed right, because many values are around 0.5, which corresponds to a moderately strong positive value. For group boundaries (F12), the prevalence of one value is due to the fact that almost all systems are based on a highly weighted indicator, which in turn consists mainly out of one highly weighted variable. Such an asymmetry is not problematic, but a particularity of the respective data set. The ecological success ranges from -0.71 to 0.99 and has an average value of 0.28. This is significantly higher than the average value of -0.17 for the CPR data. The average success of Nepalese irrigation systems is therefore higher than that of the worldwide CPR cases of the CPR database.

5.3.2 Correlations

Figure 5.3 describe the Pearson correlations of the factors in the NIIS data set. If not stated otherwise, all reported correlations are significant at the 0.001 level (positive correlations are marked clockwise, negative correlations in a counterclockwise way).

The correlation coefficients shown in Figure 5.3 are listed in the Appendix, Section 7.5.3.3.

With 156 significant correlations between success factors, the NIIS data have more correlations than the CPR data (see Appendix, Sections 7.5.2.3 and 7.5.3.3). Not surprisingly, the strongest relationship is between the size of the system (F1) and the number of actors (F7; r = 0.78).

Two success factors in particular are strongly correlated with a number of other factors: First, *characteristics of rules* (F17) correlate with social capital

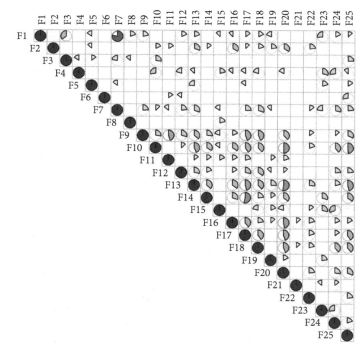

Figure 5.3 Pearson correlation coefficients between success factors of NIIS

(F9; r = 0.44), participation (F13; r = 0.46), legal certainty (F14; r = 0.53), and information (F16; r = 0.40). Second, compliance (F20) correlates with dependency on resource (F10; r = 0.50), participation (F13; r = 0.52), information (F16; r = 0.47), locally adapted rules (F17; r = 0.44), and fairness (F18; r = 0.43). Social capital is also positively correlated with resource dependency (F11; r = 0.45) and fairness (F18; r = 0.43). Finally, there is a correlation between exclusion possibilities (F22) and relationships (F23; r = 0.62).

These connections can be interpreted as follows: only within a secure legal framework and an existing high level of social capital can locally well adapted rules develop. A further condition for a well-working system is sufficient information about important information of the system (water quantity, crop yields, etc.). As with the CPR data, participation and fair rules are important for high compliance. In this data set, however, dependency on resource and information provided also play an important role. The latter were not particularly relevant in the CPR data.

For both CPR and NIIS, there are few correlations between biophysical variables (F1–F6), but many medium-sized correlations exist between success factors within the governance and actor domain.

Ecological success is directly correlated with seventeen success factors. Hence, both choice of success factors and variables is well justified. The highest influence on the output have

- dependency on resource (F10; r = 0.49),
- participation (F13; r = 0.46),
- legal certainty (F14; r = 0.35),
- locally adapted rules (F17; r = 0.41),
- regenerative capacity (F6; r = -0.27), and
- initial environmental success (F4; r = -0.12, p<0.05).

The two most important factors for the CPR data as calculated by regressions, random forests, and partly in neural networks are also correlated with success for the NIIS data set: legal certainty (F14; r = 0.35) and fairness (F18; r = 0.38).

5.3.3 Multivariate linear regressions

The regression shown in detail in Table 5.10 refers to the fully weighted and imputed NIIS data set with all variables. The results of other model variants are shown in Section 5.6.2. Model quality is shown in Table 5.11.

Table 5.10 Multivariate linear regression for the complete NIIS data set

Model MLR	Non-standardized coefficients		Standardized coefficients	T	Sig.	95.0% Confidence-interval for B		Collinearity statistics	
	Regression coefficient B	Standard error	Beta			Lower limit	Upper limit	Tolerance	VIF
(Constant)	-.144	.080		-1.801	.073	-.301	.013		
F1	.041	.056	.059	.733	.464	-.070	.152	.304	3.291
F2	.029	.058	.026	.493	.622	-.086	.143	.728	1.373
F3	.164	.047	.181	3.508	.001	.072	.256	.740	1.352
F4	.060	.034	.091	1.754	.081	-.007	.127	.733	1.365
F5	.237	.055	.219	4.275	.000	.128	.346	.749	1.336
F6	-.148	.054	-.129	-2.745	.007	-.255	-.042	.885	1.130
F7	.000	.056	.000	-.006	.995	-.111	.110	.310	3.230
F8	-.066	.078	-.042	-.852	.395	-.220	.087	.813	1.230
F9	.096	.083	.075	1.159	.248	-.067	.260	.471	2.122
F10	.310	.076	.247	4.059	.000	.159	.460	.528	1.894
F11	-.140	.051	-.157	-2.746	.006	-.240	-.040	.598	1.673
F12	.081	.104	.041	.775	.439	-.124	.286	.709	1.411
F13	.367	.093	.261	3.960	.000	.184	.549	.452	2.214

F14	.219	.093	.139	2.369	.019	.037	.401	.566	1.767
F15	.014	.060	.013	.238	.812	-.104	.133	.667	1.500
F16	-.090	.049	-.104	-1.843	.067	-.187	.006	.611	1.637
F17	.236	.095	.160	2.478	.014	.048	.423	.470	2.126
F18	.089	.059	.088	1.502	.134	-.028	.205	.565	1.769
F19	.026	.063	.022	.423	.673	-.097	.150	.742	1.348
F20	-.046	.066	-.045	-.686	.493	-.176	.085	.451	2.218
F21	-.037	.037	-.048	-1.008	.315	-.110	.036	.880	1.137
F22	.049	.036	.069	1.383	.168	-.021	.119	.795	1.257
F23	-.189	.059	-.181	-3.175	.002	-.305	-.072	.604	1.655
F24	-.072	.049	-.082	-1.457	.146	-.169	.025	.621	1.610

Table 5.11 Explanatory power of the multivariate linear regression for the complete NIIS data set

R	R^2	Adjusted R^2	Standard error of estimator	Change in R^2	Change in F	df1	df2	Sig. change in F
.731	.534	.487	.276	.534	11.365	24	238	.000

This regression shows that manageability (F5), dependency on resource (F10), and participation (F13) are important for success. Beta coefficients for F6, F11, and F23 are negative. This is plausible for regenerative ability—the slower the resource regenerates, the more important a careful handling, which may result more rapidly in sustainable rules. It is also plausible for dependency on group since in some or many cases subsistence is affected, which leads to overexploitation against better knowledge. However, such interpretations have to be considered with caution—it is easy to tell plausible "just-so" stories. Such results can only be trusted if multiple data sets and multiple quantitative methods agree, which is one reason for using three methods.

A case in point is the negative coefficient for relationships with other stakeholders (F23). Bad relationships with other groups do not make sense for improving success. Yet, one possible explanation involves the particular coding of this factor: in this data set there are practically only relations with the government. This negative correlation means that a system independent of the Department of Irrigation (DOI) is more promising than tight control, which may involve imposed, non-locally adapted solutions, corrupt and unmotivated officials, or requirements that are harmful to sustainability (Tang 1992).

In general, the model quality of a MLR on the full NIIS data is good and is equal to that of the CPR data at a respective adjusted R^2 of 0.49 (Table 5.11). The model quality of different splits is even closer together than for CPR, that is, stable between 0.40 and 0.47 (Table 5.12).

The most parsimonious model with decent explanatory power has six dependent variables and achieves an adjusted R^2 of 0.43. It includes participation (F13), regenerative capability of resource units (F6), dependency on resource (F10), manageability (F5), locally adapted rules (F17), and presence of information (F16).

Finally, Table 5.12 summarizes model performance for various splits of training and test data. As can be seen, the predictive accuracy is very similar in both the training and test data sets.

Table 5.12 Explanatory power of the multivariate linear regression for different splits of the NIIS data set

Split (training and test set)	Adjusted R^2 on the training set	Pseudo-R^2 on the test set
Relations	0.40	0.46
Success	0.42	0.44
Size	0.47	0.24
Type of management	0.42	0.36
Participation	0.41	0.44
Social capital	0.42	0.40
Revisit	0.41	0.42

Table 5.13 Explanatory power and key data of the random forests for NIIS

Split (training and test set)	% explanation of variance	Mean square deviation	Number of trees	Pseudo-R^2 on test data	Number of variables used at split
Relations	39.57	0.068	500	0.52	8
Success	38.25	0.072	1500	0.58	4
Size	44.22	0.065	500	0.29	4
Type of management	40.17	0.069	500	0.46	8
Participation	39.95	0.069	1500	0.46	8
Social capital	40.20	0.072	1000	0.52	8
Revisit	38.25	0.058	1500	0.56	4
Total data set	43.08	0.065	1500	– (due to being total data set)	8

5.3.4 Random forests

The different splits for the random forest models are very similar as well. The highest explanation of variance is 44% for the split for size (Table 5.13). The highest pseudo-R^2 on the test data set is 0.58.

Since these models are very close together, results can be considered robust. Between 38.25% (revisits) and 44.22% (size) of the variance are explained (Table

5.13). The mean value of variance explained across all seven models in Table 5.13 is 40%. This makes them comparable (as CPR) in model quality to regressions. The importance of the individual success factors is also similar (Table 5.14).

According to Table 5.14, the most important factors for success across all random forest variants, which are all imputed (NIIS unweighted, NIIS Top 3 unweighted, NIIS indicator weighting, NIIS Top 3 indicator weighting, NIIS fully

Table 5.14 Importance of success factors for random forests for NIIS

Success factor	% increase in MSE
F13 Participation	24.10
F10 Dependency on resource	16.11
F5 Manageability	12.61
F9 Social capital	10.60
F17 Characteristics of rules	10.43
F18 Fairness	10.17
F6 Regeneration of RU	10.00
F14 Legal certainty and legitimacy	6.86
F23 Relations	6.74
F20 Compliance	6.22
F24 Capabilities to adapt to change	5.78
F3 Accessibility	5.53
F8 Group composition	5.03
F11 Dependency on group	4.14
F12 Group boundaries	3.96
F7 Number of actors	3.60
F19 Control	3.33
F16 Information	3.23
F15 Administration	2.77
F1 Resource size	2.65
F22 Exclusion	2.52
F21 Conflict management	2.39
F2 Resource boundaries	1.49
F4 Ecological success at the beginning	1.03

weighted, NIIS Top 3 fully weighted, NIIS individual weight fully weighted, NIIS unweighted without revisits, NIIS unweighted without revisits Top 3), are good opportunities for participation (F13), high dependency on resource (F10), good manageability (F5), and high social capital (F9). These results are compared with CPR and IFRI in the discussion section in Chapter 6.

5.3.5 Neural networks

For the CPR data, neural networks proved to be the modeling method with the highest explanation quality. The same is true for the NIIS. Table 5.15 shows the best networks for NIIS specified according to the various splits.

The best models of neural networks range between a pseudo-R^2 of 0.36 and 0.62, with obvious differences. As with CPR, the distribution by success achieves the best model quality (62%). The best network is very generalizable since it is trained only during 75 epochs and has 8 hidden neurons. The model quality is ultimately lower than the model for the CPR data, but still very good with over 60% variance explained. The best robust net is trained only slightly longer (100 epochs), has 23 neurons in the hidden layer, and achieves a pseudo-R^2 of 0.56 (Table 5.16). The worst networks achieve a R^2 of 0.30. As with the CPR data set, neural networks are clearly superior to the other two methods. The best robust neural networks for NIIS confirm this, reaching an explanatory power of 56% for the split by success.

Table 5.15 Explanatory power and key data for the best neural networks for the different test data splits of NIIS

	Relations	Success	Size	Type of management	Participation	Social capital	Revisit
Number of neurons in hidden layer	22	8	21	10	23	27	28
Epochs	150	75	350	125	250	100	325
Algorithm	RPROP	RPROP	RPROP	RPROP	RPROP	RPROP	RPROP
Min. error	−0.44	−0.43	−0.47	−0.40	−0.49	−0.44	−0.52
Max. error	0.62	0.47	0.51	0.68	0.50	0.41	0.55
MSE	0.05	0.04	0.06	0.07	0.05	0.05	0.05
MAE	0.18	0.16	0.21	0.22	0.18	0.17	0.18
R^2	0.57	0.62	0.42	0.36	0.54	0.47	0.58

Table 5.16 Explanatory power and key data for the robust neural networks for the different NIIS test data splits

	AMIS/ FMIS (Type)	Relations	Success	Size	Intervention	Participation	Social capital
Number of neurons in hidden layer	7	11	23	21	14	14	27
Epochs	75	75	100	175	125	100	50
Algorithm	RPROP	RPROP	RPROP	RPROP	RPROP	RPROP	RPROP
Min. error	−0.38	−0.65	−0.41	−0.65	−0.59	−0.52	−0.43
Max. error	0.80	0.60	0.60	0.69	0.59	0.70	0.55
MSE	0.08	0.06	0.05	0.08	0.07	0.07	0.05
MAE	0.22	0.19	0.16	0.22	0.20	0.21	0.17
R^2	**0.30**	**0.51**	**0.56**	**0.30**	**0.45**	**0.40**	**0.43**

In addition to model quality, the importance of the individual success factors is also of interest. Table 5.17 shows the average weight over nine model variants, which are all imputed—NIIS unweighted, NIIS Top 3 unweighted, NIIS indicator weighting, NIIS Top 3 indicator weighting, NIIS fully weighted, NIIS Top 3 fully weighted, NIIS individual weight fully weighted, NIIS unweighted without revisits, NIIS unweighted without revisits Top 3.

Averaging across all splits results in the most important factor being regenerative capacity (F6) before participation (F13), followed by relationships to others (F23) and manageability (F5), monitoring (F19), and legal certainty (F14). This results in a very good agreement with the most important factors of MLR and random forests. Except for F19, the most important factors are the same as MLR.

5.3.6   Discussion

Unlike CPR, NIIS data comes from a single country, Nepal. A greater homogeneity can therefore be expected, especially as only one sector, irrigation systems, is being investigated. The descriptive statistics (Appendix, Section 7.5.3.1) clearly show that—as before with CPR—NIIS irrigation systems are small with clear boundaries. Relatively homogeneous groups act with a high social capital and locally adapted rules. They often have good opportunities to participate. Large irrigation systems administered by the government are thus in a minority. The dependent variable, ecological success, is slightly positive (mean value 0.28).

Table 5.17 Importance of success factors for neural networks for NIIS

Success factor	Mean relevance (total)	Mean standard deviation (total)
F6 Regeneration of RU	17.16	6.05
F13 Participation	16.97	6.12
F23 Relations	16.77	5.95
F10 Dependency on resource	16.29	6.18
F5 Manageability	15.87	6.14
F19 Control	14.37	6.45
F14 Legal certainty and legitimacy	14.11	6.56
F11 Dependency on group	13.12	6.48
F3 Accessibility	13.01	6.29
F17 Characteristics of rules	12.96	6.59
F12 Group boundaries	12.34	6.59
F20 Compliance	12.30	6.56
F9 Social capital	11.75	6.51
F8 Group composition	11.70	6.54
F24 Capabilities to adapt to change	11.39	6.39
F22 Exclusion	11.22	6.41
F18 Fairness	10.74	6.28
F16 Information	10.43	6.28
F21 Conflict management	10.24	6.33
F15 Administration	9.67	6.28
F4 Ecological success at the beginning	9.62	6.28
F2 Resource boundaries	9.49	6.22
F1 Resource size	9.42	6.17
F7 Number of actors	9.05	6.26

The regressions models perform well and usually range between an adjusted R^2 of 0.40 and 0.55. The standard regression on the complete, fully weighted, and imputed data achieves a R^2 of 0.49, the same as for CPR.

Correlations show a very coherent picture that is consistent with the literature. The highest correlation is to be expected between the number of actors (F7) and the resource system size (F1), which is the case for NIIS, although this was not the case for the CPR. In addition, the correlation matrix (Figure 5.3) shows a number of correlations between group and governance system factors (F9–F20). There are clear links between fair rules (F18) and adapted rules (F17) with social capital, opportunities for participation, and also among each other. Above all, a strong dependency on resource, good opportunities for participation, and the previously mentioned adapted and fair rules then seem to lead to good rule compliance (F20).

There are also strong and direct connections with success. Thus, participation possibilities and adapted rules as well as a high dependency on resource contribute directly to success (correlation coefficients each above 0.4), and also legal certainty (F14), fairness (F18), and good compliance with rules (F20) are positively correlated with success (each above 0.3). Thus, the two most important factors in CPR data (F14 and F18) are also significant for irrigation systems in Nepal—at least in the correlations. In contrast, they no longer play an outstanding role in the models.

In a comparison of model performance, regressions and random forests are comparable. While regressions lie, depending on distribution, between an adjusted R^2 of 0.29 to 0.55, random forests range between 0.40 and 0.48 of variance explained. The adjusted R^2 for the complete, fully weighted, and imputed model is 0.49 and 0.44 respectively. Again, neural networks prove to be the best method, as they reach values between 0.36 and 0.62. The best robust network is 0.56, which represents a minus of 0.13 in predictive accuracy compared to the CPR data set (0.69).

Again, it is of interest whether the three methods converge in the assessment of the most important factors. With the CPR data it is regressions and random forests that are more congruent. Here, all three methods converge in their estimates. The most important factors in the regression models are participation (F13), regenerative capability of the resource units (F6), dependency on resource (F10), and manageability (F5). These four factors alone account for 39% of the variance. The random forests also identify participation (F13), dependency on resource (F10), and manageability (F5) as relevant. They are also consistent with the neural networks, which identify regenerative capacity (F6), participation (F13), relationships with other actors (F23), dependency on resource (F10), and manageability (F5) as the most important factors. For three factors (F13, F5, and F10), this shows an almost complete agreement across methods.

The first three factors have also been considered important in the literature on irrigation systems. Thus, an almost general superiority of farmer-managed systems over government-led systems can be demonstrated (Tang 1992; Lam 1998; Frey, Villamayor-Tomas, and Theesfeld 2016). This is reflected in the success factor participation. The dependency of the crop on rainfall over the course of the year is particularly evident in regeneration capacity. Both rice and cereals—which are generally grown in Nepal and also in this dataset—are very sensitive to too little or too much water (such as extreme monsoon periods) in certain periods of the year. Moreover, as this is mainly subsistence management, the willingness to invest in repair measures of the canal system is great due to this dependency. The last factor, manageability, underlines this dependency, as it not only measures the difficulties of transporting the units, but also the predictability when resource units will be available (here: water availability in certain minimum quantities at certain times). Reliability is therefore an important factor for NIIS. As with CPR, the models explain a large part of the variance in ecological success—between 40% and 60%, which has not yet been achieved for such a broad data pool.

5.4 Results for the international forestry resources and institutions data

The results of the IFRI data analysis are also presented according to the structure used for the CPR and NIIS data. Goodness-of-fit is less satisfactory than for both CPR and NIIS, which is due to the imputation, although there is one non-imputed MLR-model with a high goodness-of-fit of 66% of variance explained (see Section 5.6.3.1). All methods agree that ecological success at the beginning (F4) is by far the most important factor for success in IFRI-forests.

5.4.1 Descriptive statistics

In the IFRI data, measures of central tendency and dispersion are consistently inconspicuous (see Appendix, Section 7.5.4 and Tables 7.9 and 7.10 for the corresponding histograms and statistical key figures). Exceptions are that some factors do not reach their possible end values, such as success at the beginning (F4), for which values only up to 0.55 are available, and resource dependency (F10) and adaptability (F24), which only reach a maximum of 0.31 and 0.61, respectively. Some mean values are shifted. The mean of F2 is shifted to the left (−0.59), that of manageability (F5) and regenerative capability (F6) to the right (0.29 and 0.27, respectively). Other success factors (F4, F10, F15, and F24) display a certain skewness and kurtosis; visually, however, their histograms show only a slightly shift. Success at the beginning (F4) is characterized by the fact that the majority of cases have the same value (0.08). The dependent variable, success, spans a range of −0.74 to 0.79 and is quite central with low variance (0.06) and a mean of 0.11.

5.4.2   Correlations

Figure 5.4 depicts the Pearson correlations for the IFRI data set. Correlations refer to the fully weighted and imputed data set (positive correlations are marked clockwise, negative correlations in a counterclockwise way).

The numerical values of this correlation matrix (Figure 5.4) are shown in the Appendix, Section 7.5.4.3.

There are 146 significant correlations within the IFRI data set (see Appendix, Section 7.5.4.3). However, most of them are weaker than NIIS and show correlation coefficients around 0.2. Particularly strong correlations exist between

- information (F16) and social capital (F9; r = 0.48),
- participation (F13; r = 0.47) and relationships (F23; r = 0.52),
- participation (F13) and social capital (F9; r = 0.41),
- relationships with others (F23) and group composition (F12; r = 0.34),
- number of actors (F7) and exclusion possibilities (F22; r = 0.33),
- manageability (F5) and accessibility (F3; r = 0.49), which is a rare relationship between two biophysical variables, and
- regenerative capacity (F6) and resource dependency (F10; r = −0.42).

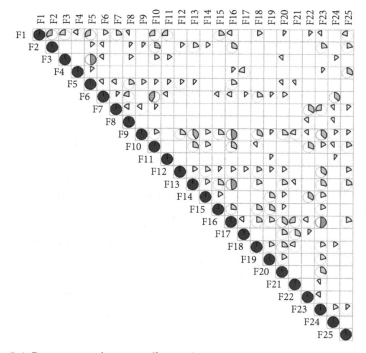

Figure 5.4 Pearson correlation coefficients between success factors of IFRI

We have already encountered correlations between participation, informa-tion, and social capital in the other data sets. These seemingly represent robust connections. At the same time, this complex of success factors has an impact on relations with others (F23). In contrast, the connection between manageability, that is, for example, sawing and transport on the one hand and accessibility on the other hand, is forestry-specific, resulting from the type of resource unit. It does not play such a big role in fisheries and irrigation systems.

Ecological success is positively correlated with ten variables between 0.12 and 0.31, which represent only moderate correlations. The strongest correlations exist with success at the beginning (F4; $r = 0.31$) and participation (F13; $r = 0.24$). Participation is also directly correlated with success in the NIIS data. Furthermore, there is a negative relationship between size (F1) and success ($r = -0.2$).

5.4.3 Multivariate linear regressions

This section describes the regression models on the IFRI data set (Table 5.18). The results of the other model variants are shown in Section 5.6.3.

Beta coefficients for IFRI are lower than for CPR and NIIS. The highest coefficients are found for success at the beginning (F4), participation (F13), followed by number of actors (F7), group boundaries (F12), and acces-sibility (F3). The only negative coefficient is resource size (F1)—too large an area seems to be associated with lack of control and lack of information about resource units. The importance of resource and group boundaries also indicates that one central problem in forest management is which ac-tors possess which rights. This is not an important factor for either CPR or NIIS, even if the causal pathway is immediately understandable and often discussed in the literature.

Regression coefficients present a coherent overall picture—from the roles of participation and success at the beginning, which is especially important in trees as slowly growing resource units, to boundary problems on several levels.

The summary in Table 5.19 reports model quality again. Despite being a co-herent model, its explanatory power is only mediocre to weak ($R^2 = 0.22$).

The six splits have almost identical explanatory power (R^2 between 0.16 and 0.20 on the training and between 0.12 and 0.27 on the test sets). A stepwise back-ward removal of success factors leads to 17 more parsimonious models with R^2 between 0.217 to 0.237. As in CPR and NIIS, model quality ranges in a narrow corridor. However, the model with the fewest influencing factors and an adjusted R^2 of 0.234 has only eight independent variables (F1, F2, F3, F4, F7, F12, F13, F14) in contrast to CPR (14) and NIIS (13).

Table 5.18 Multivariate linear regression for the complete IFRI data set

Model MLR	Non-standardized coefficients		Standardized coefficients	T	Sig.	95.0% Confidence-interval for B		Collinearity statistics	
	Regression coefficient B	Standard error	Beta			Lower limit	Upper limit	Tolerance	VIF
(Constant)	.166	.037		4.473	.000	.093	.239		
F1	−.123	.032	−.226	−3.883	.000	−.185	−.061	.568	1.762
F2	.035	.020	.084	1.697	.090	−.005	.075	.779	1.284
F3	−.064	.031	−.114	−2.063	.040	−.125	−.003	.634	1.578
F4	.511	.084	.284	6.067	.000	.346	.677	.874	1.145
F5	−.002	.036	−.002	−.042	.966	−.072	.069	.587	1.704
F6	−.054	.050	−.064	−1.090	.276	−.152	.044	.565	1.769
F7	.126	.039	.168	3.237	.001	.049	.202	.709	1.410
F8	−.013	.040	−.015	−.317	.751	−.091	.066	.810	1.234
F9	−.002	.043	−.003	−.057	.954	−.086	.081	.600	1.667
F10	.006	.089	.004	.070	.944	−.168	.181	.518	1.929
F11	−.042	.043	−.047	−.967	.334	−.127	.043	.831	1.203
F12	.072	.029	.123	2.490	.013	.015	.129	.785	1.274
F13	.147	.036	.225	4.062	.000	.076	.218	.626	1.598
F14	−.069	.044	−.082	−1.551	.122	−.156	.018	.680	1.470

F15	.024	.040	.032	.594	.553	-.055	.102	.676	1.478
F16	-.007	.045	-.012	-.160	.873	-.097	.082	.373	2.684
F17	.046	.050	.047	.921	.358	-.052	.144	.735	1.361
F18	-.035	.036	-.051	-.979	.328	-.105	.035	.711	1.406
F19	-.004	.039	-.006	-.114	.909	-.080	.072	.767	1.304
F20	.056	.046	.066	1.230	.220	-.034	.146	.672	1.488
F21	-.002	.030	-.003	-.063	.950	-.062	.058	.781	1.281
F22	-.002	.042	-.002	-.043	.966	-.085	.081	.622	1.609
F23	-.011	.061	-.011	-.178	.859	-.130	.109	.539	1.856
F24	-.001	.058	-.001	-.020	.984	-.116	.114	.730	1.371

Table 5.19 Explanatory power of the multivariate linear regression for the complete IFRI data set

R	R^2	Adjusted R^2	Standard error of estimator	Change in R^2	Change in F	df1	df2	Sig. change in F
.513	.263	.217	.218	.263	5.701	24	384	.000

Table 5.20 Explanatory power of the multivariate linear regression for the different splits of the IFRI data set

Split (training and test set)	Adjusted R^2 on the training set	Pseudo- R^2 on the test set
Ecological success	0.18	0.18
Ecological success at the beginning	0.16	0.27
Size	0.16	0.25
Country	0.20	0.15
Social capital	0.18	0.16
Revisit	0.20	0.12

The most parsimonious model with decent explanatory power, which is obtained via automatic modeling, consists of four factors—success at the beginning (F4), number of actors (F7), size (F1), and participation (F13). Its adjusted R^2 of 0.19 is just below the full model quality. The sensitivity analyses in Section 5.6 demonstrate that the low goodness-of-fit is mainly due to imputation since the best MLR-model achieves an adjusted R^2 of 0.66.

As with NIIS, this results in four critical factors for success. Both in NIIS and IFRI, opportunities to participate (F13) are crucial. For IFRI, success at the beginning (F4) and size (F1) are important as well. The differences are probably due to the differences in resource systems—forests versus irrigation. For example, size is just as important in forests as manageability and resource dependency are in irrigation systems, which are almost always subsistence farming, at least in Nepal.

Table 5.20 summarizes model qualities for different splits.

While model quality does not differ much for training data sets (0.16 to 0.20 in the adjusted R^2), test data sets already show significant differences in accuracy, with the split by success at the beginning reaching the highest performance.

5.4.4 Random forests

Since model quality of the regressions is relatively low, the next section shows whether the very robust random forests have higher explanatory power. Like regressions, the variants of the random forests are also fairly close together in their model qualities (Table 5.21). As with regressions, the split by revisit has the best performance.

The random forests are on par with regressions in terms of model quality—between 13.5% and 18.7% depending on split. The mean value of variance explained across all six model splits is 18%. For the total data set it is also 18% and the mean pseudo-R^2 on test data sets is 0.24.

This brings us to the significance of the individual factors with regard to success (Table 5.22).

The most important factors (Table 5.22) are an initial good condition of the resource (F4), a small size of the system (F1), good participation possibilities (F13), and the regenerative ability of the trees (F6). The first two success factors turn out to be particularly important, which was also true for CPR and NIIS. Thus, there is a very good agreement between factors that are significant in regressions with some changes in rank. These calculations pertain to success across all random forests (all imputed: IFRI unweighted, IFRI Top 3 unweighted, IFRI indicator weighted, IFRI Top 3 indicator weighted, IFRI fully weighted, IFRI Top 3 fully weighted, IFRI individual weights fully weighted).

Table 5.21 Explanatory power and key data of random forests for IFRI

Split (training and test set)	% explanation of variance	Mean square deviation	Number of trees	Pseudo-R^2 on test data	Number of variables used at split
Success	16.45	0.044	1500	0.18	8
Success at the beginning	13.52	0.043	1000	0.34	4
Size	14.93	0.045	1000	0.25	8
Country	14.99	0.045	500	0.28	8
Social capital	15.94	0.045	1000	0.25	4
Revisit	18.73	0.043	1000	0.16	4
Data set total	18.43	0.043	1500	– (due to being total data set)	4

Table 5.22 Importance of success factors for random forests for IFRI

Success factor	% increase in MSE
F4 Ecological success at the beginning	26.81
F1 Resource size	22.27
F13 Participation	14.62
F6 Regeneration of RU	11.65
F7 Number of actors	11.58
F20 Compliance	11.09
F2 Resource boundaries	8.71
F3 Accessibility	8.62
F9 Social capital	8.49
F22 Exclusion	8.37
F5 Manageability	8.12
F10 Dependency on resource	7.88
F15 Administration	7.69
F23 Relations	7.64
F8 Group composition	7.27
F14 Legal certainty and legitimacy	6.82
F24 Capabilities to adapt to change	6.28
F18 Fairness	6.11
F19 Control	5.15
F12 Group boundaries	4.76
F16 Information	4.71
F17 Characteristics of rules	3.88
F11 Dependency on group	3.02
F21 Conflict management	2.36

5.4.5 Neural networks

Regressions and random forests have an adjusted R^2 of around 20%. This section shows whether the methodology of the neural networks, which so far has shown the best model quality for CPR and NIIS, also provides the best explanatory power. Table 5.23 represents the best networks for IFRI in each case.

Table 5.23 Explanatory power and key data for the best neural networks for different test data splits of IFRI

	Success	Success at beginning	Size	Country	Social capital	Revisit
Number of neurons in hidden layer	17	27	18	8	26	14
Epochs	125	300	350	50	75	500
Algorithm	RPROP	RPROP	RPROP	RPROP	RPROP	RPROP
Min. error	−0.48	−0.50	−0.49	−0.50	−0.46	−0.40
Max. error	0.46	0.42	0.45	0.39	0.44	0.45
MSE	0.04	0.04	0.03	0.04	0.03	0.03
MAE	0.15	0.16	0.14	0.16	0.14	0.14
R^2	0.25	0.36	0.33	0.20	0.29	0.30

These results demonstrate, that once again, neural networks perform best, but perform considerably better on the CPR and NIIS data. Table 5.24 shows the best *robust* results for the six splits.

The neural networks, in turn, are able to clearly surpass even the random trees in explanatory power (Table 5.23). The best networks achieve a pseudo-R^2 ranging from 0.20 to 0.36. The architecture of the best net (split by success), consists of 27 neurons in the hidden layer and is trained 300 epochs using RPROP. The best robust net has been trained on the same split and has 13 neurons in the hidden layer for 250 epochs using RPROP. It reaches a R^2 of 0.30 on the test set (Table 5.24).

We now switch from model quality to the importance of individual success factors. They are presented in Table 5.25, which is sorted by relevance.

Across all splits, the importance of success factors as calculated by the networks is (Table 5.25): initial success (F4), group composition (F8), number of actors (F7), exclusion possibilities (F22), dependency on resource (F10), social capital (F9), dependency on group (F11), and the ability to adapt to changes (F24).

There is only partial agreement with the other IFRI models: only initial success (F4) and number of actors (F7) also play an important role in regressions and random forests. Accordingly, the model of neural networks cannot explain more than a third of the variance.

The weak model qualities for IFRI require an explanation given the comparatively high number of cases, the high data quality, and the large number of variables. One might speculate that this may be due to forests having by far the

Table 5.24 Explanatory power and key data for the robust neural networks for different test data splits of IFRI

	Success	Success at beginning	Size	Country	Social capital	Revisit
Number of neurons in hidden layer	14	13	13	10	7	27
Epochs	250	250	250	50	75	500
Algorithm	RPROP	RPROP	RPROP	RPROP	RPROP	RPROP
Min. error	−0.54	−0.50	−0.50	−0.49	−0.44	−0.39
Max. error	0.45	0.47	0.45	0.44	0.40	0.47
MSE	0.04	0.04	0.04	0.04	0.03	0.04
MAE	0.16	0.17	0.15	0.16	0.15	0.16
R^2	**0.16**	**0.30**	**0.24**	**0.14**	**0.22**	**0.18**

longest regeneration time of all the resource units treated here. Therefore, sustainability is extremely difficult to achieve, as both rules and actors as a rule have a much higher fluctuation until forest stands have grown back again. The great importance of the factor success at the beginning—which tries to measure the original condition of the forest—is an indication of this. Sustainable forest management may succeed in maintaining or even increasing original stands only through decades of stable and appropriate management. Especially in forests such as those in India, Nepal, Uganda, or Latin and South America, which make up the main component of IFRI data, however, the demand for wood is higher by far than the regrowth. Thus, pressure on resource units may be very high despite sustainable rules.

5.4.6 Discussion

Unlike the NIIS data, which come from only one country, the IFRI data set stems from thirteen countries. Nevertheless, their statistical values are more homogeneous. Almost all success factors have their mean value close to zero. We must therefore assume that both small and large systems, socially closely and very loosely connected groups, and both highly independent and government-led systems coexist alongside one another.

One exception to this homogeneity is the factor resource boundaries (F2)—its mean value of -0.59 is negative, although this is quite plausible for forests.

Table 5.25 Importance of success factors for neural networks for IFRI

Success factor	Mean relevance (total)	Mean standard deviation (total)
F4 Ecological success at the beginning	20.23	4.57
F8 Group composition	14.98	6.40
F7 Number of actors	14.80	6.31
F22 Exclusion	14.33	6.35
F10 Dependency on resource	14.33	6.52
F9 Social capital	13.77	6.39
F11 Dependency on group	13.15	6.54
F24 Capabilities to adapt to change	12.61	6.75
F15 Administration	12.60	6.50
F18 Fairness	12.57	6.35
F14 Legal certainty and legitimacy	12.53	6.38
F3 Accessibility	12.37	6.49
F23 Relations	11.98	6.35
F21 Conflict management	11.94	6.43
F13 Participation	11.92	6.27
F1 Resource size	11.91	6.36
F5 Manageability	11.35	6.38
F6 Regeneration of RU	11.32	6.59
F17 Characteristics of rules	11.19	6.44
F19 Control	10.89	6.36
F12 Group boundaries	10.77	6.39
F16 Information	10.42	6.39
F20 Compliance	10.22	6.43
F2 Resource boundaries	7.83	5.88

The extremely important factor initial ecological success (F4) is also centered in the middle and at its maximum is no higher than 0.55; the dependent variable, ecological success, however, reaches a maximum of 0.79.

IFRI forests show approximately the same number of correlations as CPR (156) and NIIS (132), namely 146. One difference is that correlations are much weaker on average and range between 0.1 and 0.2. This finding confirms the heterogeneity of the systems, which are located on different continents and work with different governance systems.

Like in the other data sets, however, there is a network of mutually positive influencing factors: a high level of social capital goes hand in hand with good opportunities for participation and a functioning flow of information. This in turn has a positive impact on relations with other actors. By contrast, the size of the forest and the number of actors are only moderately correlated (r = 0.23)—this relationship is much more pronounced in irrigation systems. Again, there are many positive correlations between group characteristics and the governance systems designed by these groups.

However, in contrast to CPR and NIIS, forests have an important direct relationship between biophysical factors and ecological success: success at the beginning (F4) and actual success at the end of an observation period are closely related. Other characteristics of the resource system are also relevant: a large number of actors is positive for success. Forest units that are hardly accessible increase the chance of success.

For forests, too, high participation opportunities, a good flow of information, and good relations with other actors have a positive influence on success. This is particularly plausible for forests, which can be very complex. In forest management, for example, it is not uncommon for several groups involved to have to agree on rules. Typically, the user group itself has to communicate with a superior forest authority and other user groups that harvest other resources in the same forest. Such management takes place at several levels, the forest level, regional level, and national forest directives.

This in turn suggests taking a closer look at conflict management (F20) in forests. It turns out that high social capital (F9), good communication (F16), and fair rules (F18) go hand in hand with considerably fewer conflicts. If rules are also well adapted (F17), fewer conflicts are to be expected. Prerequisites for good relations with other groups and the state are above all a good flow of information and communication (F16), a high level of legal security (F14), and clear limits for group (F12) and resource (F2). These are the starting points for NGOs, for example, to achieve a better networking of forest communities, to promote the acceptance of external experts or to encourage group members to learn new technologies.

There is agreement on the most important factor for all three methods, initial ecological success (F4). This may be due to very slow growth rates in forests. Thus, it is extremely difficult to achieve fundamental changes. That is why the initial condition is so important. If changes are to succeed, regression models and random forests agree that participation possibilities (F13) and the size of the system (F1 or F7) are relevant. However, it should be noted that overall model performance is not as high as for CPR and NIIS and explains only 15% to 20% of the variance. With a variance explanation of about 30%, neural networks perform better, but they put several other factors into focus: besides initial success (F4) and number of actors (F7) that also appear in the other models, group composition (F8), and exclusion possibilities of third parties (F22) as well as the dependency on resource (F10) and available social capital (F9) are emphasized as particularly relevant.

The emerging picture of forest management is that of a very complex system that depends heavily on biophysical factors that are difficult to measure or even to influence positively. In extreme cases it may be clear that harvesting is not sustainable. However, due to the many interrelated factors that would need to be changed, a move towards greater sustainability in many forests is often practically impossible. In addition, there are complex interactions between many interest groups. Moreover, rules are difficult to enforce due to a lack of information flow.

5.5 Results for a combined full model

Do success factors necessarily have to be sector-specific? This question is both controversially discussed in the literature and forms the basis of one of the hypotheses of this analysis. Or, are there on the contrary success factors, for example, monitoring, that are relevant for ecological success *independently* of the particular resource system type. In order to be able to answer this question quantitatively for the first time, an overall data set is formed out of all three data sets—CPR, NIIS, and IFRI. As expected, more pronounced sectoral differences disappear, and no single factor is decisive for ecological success. Model quality is medium to good (44% of variance explained).

5.5.1 Descriptive statistics

The descriptive statistics of the combined data set are again inconspicuous (see the Appendix, Section 7.5.5 and Tables 7.11 and 7.12 for the corresponding histograms and statistical key figures). Due to the origin of the data from various sources and the high number of cases (n = 794), it is not surprising that almost all factors span the entire range from –1 to 1. The minimum mean value is –0.19 (adaptability, F24), the maximum 0.29 (resource size, F1), but the means for all factors are relatively close around 0. There is no factor that has a particularly high kurtosis or skewness. The dependent variable, success, has values from –0.74 to 0.99, a mean of 0.13 and a variance of 0.11. The following section (Section 5.5.2) shows the correlations for the imputed and fully weighted overall model.

5.5.2 Correlations

As with the individual data sets, Figure 5.5 shows the Pearson correlations of the factors in the overall model (positive correlations are marked clockwise, negative correlations in a counterclockwise way).

There are many correlations (243), mostly of medium strength (between 0.15 and 0.30). The correlation matrix of the combined data set in numerical form can be found in the Appendix, Section 7.5.5.3. The data of the individual data sets show similar tendencies and reinforce each other. Some stand out and will be discussed subsequently. Clear resource boundaries (F2) correlate positively with a small number of actors (F7; r = 0.45), with group composition (F8; r = 0.59), and with social capital (F9; r = 0.48). There is an expected relationship between number of actors (F7) and size (F1; r = 0.46), but also with social capital (F9; r = 0.40). Good participation opportunities (F13) are linked to high social capital (F9; r = 0.47), which in turn promotes fairness (F18; r = 0.44) and is associated with a higher dependency on group (F11; r = 0.53). Group composition is related to social capital (F9; r = 0.45) and to resource dependence (F10; r = 0.51).

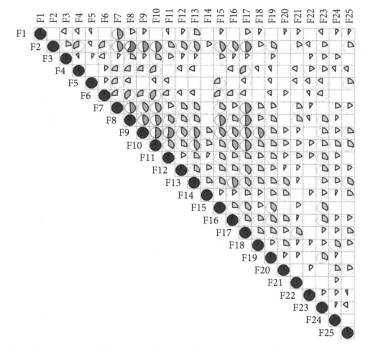

Figure 5.5 Pearson correlation coefficients between success factors of the combined model

Locally adapted rules also correlate with many other factors: with clear boundaries (F2; r = 0.49), social capital (F9; r = 0.45), dependency on resource (F10; r = 0.48), group composition (F8; r = 0.50), and number of actors (F7; r = 0.53).

These correlations reflect very well the correlations known from the literature (see Section 5.1.5). Social factors in particular—as we have already seen in the other data sets—interlock and form a complex in which it is often not completely clear what is cause and what is effect: do fair initial rules lead to greater group cohesion, or does a homogeneous group with high social capital tend to have fairer rules? Apart from these questions about cause and effect, the kind of feedback loops between factors have not been studied in a quantitative analysis.

Ecological success is directly positively correlated with thirteen factors. However, these correlations range between 0.08 and 0.31 (see the Appendix, Section 7.5.5.3), so they are not very pronounced (but all are highly significant at $p < 0.001$, expect F22, which is $p < 0.05$):

- participation opportunities (F13; r = 0.31),
- dependency on resource (F10; r = 0.29),
- locally well adapted rules (F17; r = 0.24),
- clear resource boundaries (F2; r = 0.23),
- good manageability (F5; r = 0.23),
- high social capital (F9; r = 0.23),
- legal certainty (F14; r = 0.23),
- good compliance (F20; r = 0.19),
- fair rules (F18; r = 0.18),
- success at the beginning (F4; r = −0.15),
- exclusion possibilities (F22; r = −0.07), and
- relationships (F23; r = −0.15).

5.5.3 Multivariate linear regressions

The regression shown in detail in Table 5.26 refers, as the other data sets, to the fully weighted and imputed overall model with all variables.

The model summary in Table 5.27 shows further important characteristics of the regression in addition to model quality (adjusted R^2).

The linear regression for the entire model with all cases (n = 794) achieves an adjusted R^2 of 0.33 (Tables 46 and 47). This average goodness-of-fit and the high number of factors (7) in the most parsimonious, still satisfactory model clearly prove that the configurations of success factors are reasonably resource-specific—there is regrettably no magic formula for all SES examined here (fish, forests, irrigation). This result confirms previous studies that have stated that

Table 5.26 Multivariate linear regression for the combined data set

Model MLR	Non-standardized coefficients		Standardized coefficients	T	Sig.	95.0% Confidence-interval for B		Collinearity statistics	
	Regression coefficient B	Standard error	Beta			Lower limit	Upper limit	Tolerance	VIF
(Constant)	0.083	0.024		3.473	0.001	0.036	0.130		
F1	-0.026	0.023	-0.044	-1.158	0.247	-0.071	0.018	0.598	1.673
F2	0.070	0.020	0.154	3.544	0.000	0.031	0.109	0.451	2.220
F3	-0.035	0.024	-0.048	-1.422	0.156	-0.082	0.013	0.761	1.314
F4	0.059	0.026	0.077	2.249	0.025	0.007	0.110	0.719	1.391
F5	0.175	0.025	0.227	6.910	0.000	0.125	0.225	0.791	1.265
F6	-0.059	0.030	-0.067	-1.954	0.051	-0.118	0.000	0.714	1.400
F7	0.061	0.026	0.105	2.399	0.017	0.011	0.112	0.448	2.233
F8	-0.130	0.033	-0.176	-3.982	0.000	-0.193	-0.066	0.436	2.293
F9	0.029	0.032	0.041	0.899	0.369	-0.035	0.093	0.407	2.454
F10	0.242	0.047	0.219	5.119	0.000	0.149	0.335	0.465	2.149
F11	-0.072	0.029	-0.092	-2.440	0.015	-0.129	-0.014	0.593	1.686
F12	0.020	0.030	0.022	0.649	0.517	-0.040	0.079	0.730	1.370
F13	0.171	0.034	0.198	5.009	0.000	0.104	0.238	0.546	1.832

F14	0.139	0.041	0.112	3.420	0.001	0.059	0.219	0.788	1.269
F15	-0.058	0.027	-0.082	-2.150	0.032	-0.111	-0.005	0.589	1.699
F16	-0.040	0.028	-0.054	-1.432	0.153	-0.095	0.015	0.597	1.676
F17	0.111	0.038	0.119	2.922	0.004	0.037	0.186	0.513	1.950
F18	0.071	0.028	0.088	2.539	0.011	0.016	0.126	0.704	1.420
F19	-0.031	0.032	-0.033	-0.971	0.332	-0.095	0.032	0.744	1.344
F20	0.077	0.034	0.081	2.274	0.023	0.010	0.143	0.674	1.483
F21	-0.036	0.022	-0.053	-1.639	0.102	-0.079	0.007	0.829	1.206
F22	-0.053	0.024	-0.073	-2.193	0.029	-0.100	-0.006	0.777	1.287
F23	-0.186	0.036	-0.188	-5.124	0.000	-0.257	-0.115	0.632	1.583
F24	-0.043	0.031	-0.047	-1.409	0.159	-0.103	0.017	0.776	1.289

Table 5.27 Explanatory power of the multivariate linear regression for the complete comprehensive data set

R	R^2	Adjusted R^2	Standard error of estimator	Change in R^2	Change in F	df1	df2	Sig. change in F
.588	.346	.325	.271	.346	16.932	24	769	.000

Table 5.28 Explanatory power of the multivariate linear regression for different splits of the comprehensive data set

Split (training and test set)	Adjusted R^2 on the training set	Pseudo-R^2 on the test set
Relations	0.19	0.28
Ecological success	0.20	0.22
Success at the beginning	0.23	0.12
Size	0.20	0.22
Social capital	0.19	0.26

there is no panacea (Ostrom, Janssen, and Anderies 2007; Meinzen-Dick 2007). The individual splits are closely together in their model quality (adjusted R^2 between 0.19 and 0.23) and achieve a pseudo-R^2 between 0.12 and 0.28 on the test data (Table 5.28).

Some variables prove to be more important for success than others (Table 5.26): dependency on resource (F10), relationships with others (F23), manageability (F5), participation (F13), group composition (F8), and clear resource boundaries (F2). With this model, an adjusted R^2 of 0.27 is achieved. However, a model with five factors (F5, F10, F13, and F23) still achieves an adjusted R^2 of 0.24.

If the data set is divided into training and test data set, the predictive accuracy on the test data for different splits is very similar and ranges between 0.19 and 0.26 only with the exception of one outlier (success at the beginning with 0.12).

5.5.4 Random forests

Machine learning algorithms usually perform better on larger data sets. Hence, it is to be expected that both random forests and neural networks will achieve better results on the combined data set (n = 794). However, this is not the case which is a first suggestion that success factors are mostly sector-specific.

As usual, the different models of random forests show very similar results. The highest goodness-of-fit is 38% for the split by success at the beginning (Table 5.29). The highest pseudo-R^2 on the test data set is 0.46 for the split by success.

As Table 5.29 demonstrates, random forests are superior to MLR in their predictive accuracy. Depending on the split, they explain between 32% and 38% of the variance. On the test data, the range is between 0.30 and 0.46 (pseudo-R^2). The mean value of variance explanation across all five models in Table 5.29, divided into training and test data set, is 35%, for the overall model it is 38%. The average pseudo-R^2 on test data sets 41%.

Of particular interest in the overall model is the order in which the factors are important for ecological success: Which factors prove to be robust across all data? It is to be expected that with a larger data set that contains more heterogeneous data, the importance of each individual factor is smoothed out and expected to decrease. Table 5.30 sorts the factors in descending order of importance.

Thus, participation (F13), manageability (F5), and clear resource limits (F2), followed by dependency on resource (F10) are the most relevant success factors, which means that there is an almost perfect match with the regression models and neural networks for all important factors (F10, F5, F13, and F23).

Table 5.29 Explanatory power and key data of random forests for the combined model

Split (training and test set)	% explanation of variance	Mean square deviation	Number of trees	Pseudo-R^2 on test data	Number of variables used at split
Relations	33.95	0.133	500	0.45	8
Success	31.98	0.134	1000	0.46	8
Success at beginning	38.08	0.125	500	0.30	8
Size	35.44	0.124	1500	0.42	8
Social capital	33.08	0.129	1500	0.44	8
Total data set	37.57	0.122	500	– (due to being total data set)	8

Table 5.30 Importance of success factors for random forests for the combined model

Success factor	% increase in MSE
F13 Participation	36.80
F5 Manageability	28.54
F2 Resource boundaries	27.45
F10 Dependency on resource	24.50
F23 Relations	23.59
F8 Group composition	23.45
F17 Characteristics of rules	20.80
F4 Ecological success at the beginning	19.78
F11 Dependency on group	17.83
F14 Legal certainty and legitimacy	17.39
F7 Number of actors	16.88
F24 Capabilities to adapt to change	15.00
F18 Fairness	14.50
F9 Social capital	14.30
F6 Regeneration of RU	13.73
F15 Administration	13.08
F20 Compliance	12.62
F22 Exclusion	12.53
F1 Resource size	11.42
F16 Information	11.20
F12 Group boundaries	10.56
F21 Conflict management	10.28
F3 Accessibility	9.40
F19 Control	8.00

5.5.5 Neural networks

Neural networks, even more than random forests, require large training sets for good model results. These results are again listed according to the different splits in Table 5.31.

Table 5.31 Explanatory power and key data for the best neural networks for different test data splits of the combined model

	Relations	Size	Social capital	Success	Success at beginning
Number of neurons in hidden layer	29	9	19	24	25
Epochs	150	350	275	325	250
Algorithm	RPROP	RPROP	RPROP	RPROP	RPROP
Min. error	−0.79	−1.07	−1.11	−1.03	−1.07
Max. error	0.75	0.87	0.76	1.13	0.85
MSE	0.10	0.13	0.12	0.11	0.13
MAE	0.26	0.29	0.27	0.26	0.28
R^2	0.42	0.38	0.42	0.44	0.24

Table 5.32 Explanatory power and key data for the robust neural networks for different test data splits of the combined model

	Relations	Size	Social capital	Success	Success at beginning
Number of neurons in hidden layer	15	12	6	23	14
Epochs	175	325	200	250	150
Algorithm	RPROP	RPROP	RPROP	RPROP	RPROP
Min. error	−0.85	−1.12	−1.17	−1.18	−1.08
Max. error	0.76	0.90	0.81	1.14	1.09
MSE	0.11	0.14	0.12	0.12	0.14
MAE	0.27	0.30	0.28	0.27	0.29
R^2	0.35	0.34	0.39	0.40	0.17

The best network achieves an R^2 of 0.44 on the test data sets. Similarly, in Table 5.32, the best robust networks are hardly worse in goodness-of-fit—a split by success reaches an R^2 of 0.40.

With a pseudo-R^2 of 0.44, the best artificial neural networks are on par with random trees (0.46, Table 5.29). The best robust network achieves an R^2 of 0.40. Compared to the outstanding model quality on the CPR data set, this is not as good, but significantly better compared to the IFRI data (0.40 compared to 0.30).

Finally, individual factor importance is also calculated for neural networks in Table 5.33.

Table 5.33 Importance of success factors for neural networks for the combined model

Success factor	Mean relevance (total)	Mean standard deviation (total)
F4 Ecological success at the beginning	81.49	6.37
F10 Dependency on resource	77.48	6.48
F5 Manageability	76.06	6.51
F14 Legal certainty and legitimacy	75.54	6.39
F9 Social capital	72.89	6.77
F22 Exclusion	71.78	6.87
F13 Participation	71.53	6.67
F8 Group composition	68.28	6.80
F23 Relations	65.75	6.84
F11 Dependency on group	65.31	6.76
F7 Number of actors	63.06	6.83
F18 Fairness	62.93	6.75
F15 Administration	60.38	6.63
F24 Capabilities to adapt to change	59.22	6.79
F21 Conflict management	58.00	6.65
F3 Accessibility	57.65	6.59
F6 Regeneration of RU	57.58	6.60
F17 Characteristics of rules	54.45	6.55
F20 Compliance	53.59	6.58
F1 Resource size	52.90	6.44
F12 Group boundaries	52.67	6.45
F19 Control	51.88	6.71
F2 Resource boundaries	45.72	6.42
F16 Information	43.84	6.29

Although the number of cases in the overall model is much higher than in the individual models, the calculations in Table 5.33 are based on just two data variants and may thus not be as robust as the CPR and IFRI (based on 7 variants) or NIIS (9).

Still, the splits predict that success at the beginning (F4), dependency on resource (F10), manageability (F5), and legal certainty (F14) are the most relevant factors. Across methods, neural networks agree with regressions and random forests that factors F5, F10, and F13 are important for ecological success.

5.5.6 Discussion

Due to the fact that the overall model combines three data sets, there are no more outliers in the descriptive statistics for all twenty-four factors and ecological success. All factors are centered on zero, almost always exploit the minimum and maximum perfectly, and show no large kurtosis or skewness. Interestingly, however, twenty-one of twenty-four factors including success have shifted their mean value slightly to the right. This suggests that case selection is slightly shifted towards successful systems, presumably simply because some of the unsuccessful systems no longer exist. This sample selection bias may be due to the fact that none of the data sets was randomized regarding case selection.

In the combined model, the network of positively correlated group and governance system factors is also strong. Again, participation possibilities (F13), social capital (F9), adapted rules (F17), and dependency on resource (F10) are connected closely. It is impossible to determine what is cause and what is effect. In addition, there are many correlations with clear resource boundaries, such as a smaller number of actors (F7), a homogeneous group composition (F8), and a good flow of information (F16). Relations with other actors (F23) are also closely linked to other factors, for example with fair rules (F18) or good provision of information (F16).

This in turn suggests the well-known picture of a small, closely cooperating, ethnically, economically, or religiously homogeneous community that is dependent on the resource on a subsistence basis and has managed to create established norms through sensible rules that (almost) everyone in the small community adheres to—even without government intervention.

Model quality of the overall data set is satisfactory for the regressions with 33% of the explained variance, while the best robust neural networks reach 40% and random forests 38%. However, compared to the sector-specific models for CPR, NIIS, and IFRI, the overall model performs somewhat worse. This confirms the suspicion often expressed in the literature that there is no panacea (Ostrom, Janssen, and Anderies 2007; Meinzen-Dick 2007). A further corroboration for this hypothesis stems from the finding that a multitude of factors is necessary for models to reach an adequate explanatory quality—one or two factors are not sufficient.

Participation (F13) and manageability (F5) are particularly important in the overall model. Clear resource boundaries (F2), dependency on resource (F10), and relationships with other actors (F23) are also important in random forests (Table 5.30). This is largely confirmed by neural networks: in addition to these factors, good initial ecological success is relevant (F4), while F10 and F5 are confirmed as equally important and participation (F13) loses some influence (Table 5.33). In these models, clear resource boundaries (F2) do not play any role.

5.6 Robustness and sensitivity analyses

An important complement to the models above are sensitivity analyses, that is, consistency and robustness of results. Sensitivity analyses deal with the question of whether results are robust against parameter changes or data transformation procedures. A number of steps have been taken to ensure that results can be trusted. This section is about describing various individual sensitivity analyses that provide additional transparency.

First, as demonstrated, a separate model for each critical data transformation decision—imputation, weighting, number of variables used, and method choice—has been calculated. Thus, for example, weighting the variables and indicators could be seen as biasing the data by subjective opinion (see Sections 3.5.3 and 3.5.4). Although the weighting has been performed independently by three people, and although the degree of agreement is very high and satisfactory (Krippendorff's Alpha is $\alpha = 0.901$ for NIIS), a subjective influence nevertheless remains.

Thus, for each data set (CPR, NIIS, IFRI) and the combination of them a model is available for comparisons—imputed and non-imputed, weighted and unweighted, with all or only the three most important variables. Each of these variants is modeled with regressions, random forests, and neural networks. The disadvantage is that very many models have to be tested due to the permutations of these combinations. The advantage outweighs this, however, since a very robust assessment of results becomes possible. The results for all combinations are presented in Sections 5.6.1, 5.6.2, and 5.6.3.

Second, reproducibility and robustness are checked by repeating each run of neural networks three times. Only if all three results of goodness-of-fit are sufficiently close together (deviation of variance =<0.08), a model is evaluated as replicated. Only then the respective network configuration is included in the selection for the best robust nets. Differences between runs can be explained by the fact that the learning algorithms of the neural networks use pseudo-random numbers as seeds. In general, the average deviation is very low.

Third, a general check of neural network error rates across all data set combinations is performed. Three statements can be made. No model has a particularly high error rate on the unknown test data—no result is worse than 0.32 (mean absolute error, MAE) or 0.16 (mean squared error, MSE) in any configuration. Table 5.34 provides an overview. The range of error rates is small, which speaks for the robustness. The average of the lowest errors for the best networks of all data sets is 0.18 (MAE) and 0.06 (MSE). The average of the variance explained is 44% (Table 5.34, second column).

The average MAE across all configurations and data sets distinguishing between fully weighted and unweighted data reveals that errors of the unweighted variants are smaller: the mean MAE value for CPR is 0.154 for fully weighted data, 0.143 for unweighted data; the values for NIIS are 0.203 and 0.185; for IFRI 0.190 and 0.168. However, this test averages across all configurations—the informational value is thus limited. There is no clear trend in the weighting of the model variants themselves—sometimes unweighted models are more accurate, sometimes not—which in turn speaks in favor of this precautionary measure chosen. Ultimately, only the best and most robust models are of interest for further analyses.

Fourth, a model calibration check is performed. By manually setting the success factors to their minima, their maxima, and to a neutral value of zero in the trained neural networks, predictions can be tested. Take as example the model for NIIS. If all twenty-four success factors are set manually to their minima of −1.0 in the trained model, this results in a prognosis of −0.99 for ecological success. If all factors are set to their maxima of 1.0, an ecological success of 0.85 is predicted. At consistently neutral values (all factors at 0.0), the network predicts −0.29. Hence, the models are calibrated well, the range of possible success and failure is exploited and meets expectations. In addition, neutral values do not seem to be sufficient to achieve positive values for ecological success, which is a promising hypothesis to test with other data sets.

Fifth, the model calibration checks just described are extended to test for internal consistency. A group of factors (resource, resource units, actors, control system, external influences) is set to their maximum (1.0) and all other factors to their minimum (−1.0). Then, the prediction for ecological success is determined. If the other factors are now set from their minima to neutral values (0.0), this should result in a more positive prediction. This is indeed the case for all factor groups, indicating internally consistent models. If the other factors are set to their minima, no predicted value is better than −0.9. If they are neutral, they range between 0.1 and 0.3. The only exception is the factor group external environment, where a negative result is also predicted for neutral values.

Table 5.34 Mean absolute error and mean standard error for all data sets

Data	Variant	R^2	Lowest MSE	Lowest MAE
CPR	Not weighted, imputed	0.562	0.023	0.112
NIIS	Not weighted, imputed	0.618	0.037	0.156
IFRI	Not weighted, imputed	0.249	0.037	0.154
CPR	Top 3, not weighted, imputed	0.753	0.015	0.100
NIIS	Top 3, not weighted, imputed	0.523	0.045	0.163
IFRI	Top 3, not weighted, imputed	0.058	0.162	0.319
CPR	Indicator weighting, imputed	0.457	0.034	0.146
NIIS	Indicator weighting, imputed	0.605	0.056	0.178
IFRI	Indicator weighting, imputed	0.136	0.061	0.194
CPR	Top 3, indicator weighting, imputed	0.704	0.018	0.098
NIIS	Top 3, indicator weighting, imputed	0.352	0.089	0.238
IFRI	Top 3, indicator weighting, imputed	0.303	0.158	0.316
CPR	Fully weighted, imputed	0.679	0.015	0.088
NIIS	Fully weighted, imputed	0.611	0.055	0.188
IFRI	Fully weighted, imputed	0.146	0.061	0.188
CPR	Top 3, fully weighted, imputed	0.710	0.017	0.098
NIIS	Top 3, fully weighted, imputed	0.431	0.058	0.186
IFRI	Top 3, fully weighted, imputed	0.245	0.148	0.315
CPR	Individual weighting, fully weighted, imputed	0.364	0.044	0.165
NIIS	Individual weighting, fully weighted, imputed	0.212	0.081	0.217
IFRI	Individual weighting, fully weighted, imputed	0.501	0.063	0.202
Mean		**0.439**	**0.061**	**0.182**

However, it is close to 0 and hardly varies at all. These results are also confirmed by theoretical considerations. Many case studies stress that there is no single group of success factors that can be isolated (for a meta-analysis see Cox, Arnold, and Villamayor Tomas 2010). The described examination of internal consistency confirms this; if all factors except one group of factors are set to their minima, a massive failure is predicted for the ecological success in SES.

Sixth, calculation of factor significance of neural networks is in itself a sensitivity analysis as well since predictive accuracy is averaged across tens of thousands of repetitions. The many different network configurations ensure that individual results are averaged out and that the actual factor significance for success is reasonably robust. However, the standard deviation for each factor remains relatively high—uncertainties remain.

Finally, this section report the systematic permutations of the three data sets. Comparing differences in model quality between methods and data transformation procedures emphasize how sensitive models are. The many different model variants increase robustness and transparency by showing the corridor of model explanatory power. Furthermore, the respective best model can be picked from these overviews. The following parameters are changed:

- instead of all variables only the three most important per factor are considered ("Top 3")
- missing values are not imputed, but remain empty (this only applies to regressions, since both neural networks and random forests require complete input)
- weighting is changed
  o unweighted raw data: neither variables nor indicators are weighted
  o pure indicator weighting: variables are not weighted, indicators are weighted
  o change of weighting: variables are weighted, so are indicators, but a different weighting is used (instead of the average weighting of three independent persons, weighting of an individual is used).

The combinations of these three decisions (all variables/Top 3 variables; imputed data/not imputed data; unweighted, indicator-weighted, fully weighted data) result in the eight data set variants shown throughout the following sections (Sections 5.6.1–5.6.3).

5.6.1 Common-pool resource data

5.6.1.1 Multivariate linear regressions

Table 5.35—as do all following tables in the following sections (Sections 5.6.1–5.6.3)—shows the goodness-of-fit for the two different data variants (all variables/Top 3 variables) for all weightings and imputations. A dash means that a variant has not been calculated, for example, because imputation is necessary for random forests and neural networks. A value in bold means that this is the best model for the particular data set and column.

Table 5.35 Goodness-of-fit (adjusted R^2) of regressions for CPR and CPR Top 3 per weighting and imputation on both full data and test set

Description of parameters varied	CPR	CPR Top 3	CPR (test)	CPR Top 3 (test)
Fully weighted (Var + Ind +), imputed	41	49	50	58
Fully weighted (Var + Ind +), not imputed	**63**	62	13	60
Fully weighted (Var + Ind +), imputed, individual weighting	44	—	11	—
Fully weighted (Var + Ind +), not imputed, individual weighting	61	—	26	—
Indicator weighting (Ind +), imputed	42	40	9	30
Indicator weighting (Ind +), not imputed	57	61	21	23
Not weighted, imputed	41	40	15	30
Not weighted, not imputed	**63**	61	**81**	23

The strongest model by far (goodness-of-fit of 81%) for the test data sets is neither weighted nor imputed and contains all variables. The other non-imputed models here also tend to be better.

5.6.1.2 Random forests

As with regressions, Table 5.36 is an overview of random forest results.

These variants show the robustness of random forests, since results hardly vary between weightings and the number of variables (all vs. Top 3). The best model on the training data is unweighted, imputed, and includes all variables (61% of variance explained). For the test set, the most predictive model is fully weighted, imputed, and includes all variables (64%).

5.6.1.3 Neural networks

The results for the neural networks are presented in Table 5.37.

Neural network models perform consistently better using only three variables. The best model is unweighted, non-imputed, and uses only the three best variables (goodness-of-fit = 75%). On the test data, the best model is fully weighted, imputed, and contains only the best three variables (69%).

Table 5.36 Goodness-of-fit (adjusted R^2) of random forests for CPR and CPR Top 3 per weighting and imputation on both full data and test set

Description of parameters varied	CPR	CPR Top 3	CPR (test)	CPR Top 3 (test)
Fully weighted (Var + Ind +), imputed	54	48	**64**	46
Fully weighted (Var + Ind +), imputed, individual weighting	53	—	34	—
Indicator weighting (Ind +), imputed	53	49	26	45
Not weighted, imputed	**61**	48	54	46

Table 5.37 Goodness-of-fit (adjusted R^2) of best neural networks for CPR and CPR Top 3 per weighting and imputation on both full data and test set

Description of parameters varied	CPR	CPR Top 3	CPR (test)	CPR Top 3 (test)
Fully weighted (Var + Ind +), imputed	71	72	61	**69**
Fully weighted (Var + Ind +), imputed, individual weighting	36	—	27	—
Indicator weighting (Ind +), imputed	46	72	42	65
Not weighted, imputed	56	**75**	50	61

As is the case with regressions and random forests, indicator-weighted data tend to be worse than both fully weighted and unweighted variants. The goodness-of-fit of Top 3 models is basically the same if weightings are changed.

5.6.2 Nepal irrigation institution study data

5.6.2.1 Multivariate linear regressions
As with the CPR data, results for the NIIS are presented according to the three statistical methods (Table 5.38).

Results are consistently similar and prove to be very robust against all variations. The best model is unweighted—neither imputation nor the number of variables included play a role (52% explanation of variance).

Table 5.38 Goodness-of-fit (adjusted R^2) of regressions for NIIS and NIIS Top 3 per weighting and imputation on both full data and test set

Description of parameters varied	NIIS	NIIS Top 3	NIIS (test)	NIIS Top 3 (test)
Fully weighted (Var + Ind +), imputed	49	41	54	33
Fully weighted (Var + Ind +), not imputed	44	41	29	35
Fully weighted (Var + Ind +), imputed, individual weighting	39	—	40	—
Fully weighted (Var + Ind +), not imputed, individual weighting	41	—	44	—
Indicator weighting (Ind +), imputed	49	46	54	35
Indicator weighting (Ind +), not imputed	46	51	47	**55**
Not weighted, imputed	**52**	44	52	36
Not weighted, not imputed	49	**52**	51	30
Not weighted, imputed (no revisit, n = 244)	46	46	45	33
Not weighted, not imputed (no revisit, n = 244)	46	**52**	47	47

On the test data, regressions on the NIIS for the first time show that the strongest model is among the indicator-weighted data sets—it is not imputed and contains only the three most important variables (goodness-of-fit = 55%). Imputation has a rather positive effect on Top 3 data and on data sets with all variables. In general, the latter perform slightly better than Top 3 data sets.

5.6.2.2 Random forests
Table 5.39 shows the results of random forests for the NIIS.

Again, these models hardly differ from each other and are very robust compared to the changes made. The best model is not weighted, imputed, and contains only the three most important variables per factor (goodness-of-fit = 49%). On the test set data, the strongest model is indicator-weighted, imputed, and contains all variables (58% explanation of variance). There are no other trends.

Table 5.39 Goodness-of-fit (adjusted R^2) of random forests for NIIS and NIIS Top 3 per weighting and imputation on both full data and test set

Description of parameters varied	NIIS	NIIS Top 3	NIIS (test)	NIIS Top 3 (test)
Fully weighted (Var + Ind +), imputed	44	44	57	52
Fully weighted (Var + Ind +), imputed, individual weighting	40	—	42	—
Indicator weighting (Ind +), imputed	44	47	**58**	48
Not weighted, imputed	45	**49**	42	52
Not weighted, imputed (no revisit, n = 244)	42	45	46	55

5.6.2.3 Neural networks

Table 5.40 shows the results for model accuracy of neural networks.

Here, data sets with all variables are superior. Weighting seems to have no consistent influence. The best model is unweighted, imputed, and includes all variables (goodness-of-fit = 62%). The best robust model on the test data is also unweighted, imputed, and includes all variables (goodness-of-fit = 59%).

5.6.3 International forestry resources and institutions data

5.6.3.1 Multivariate linear regressions

As with CPR and NIIS data, this section presents the results for the IFRI data. Table 5.41 shows the goodness-of-fit for the different variants.

The regression variants for IFRI show strong differences in model quality. The strongest model is also the "most parsimonious": it is neither weighted nor imputed, nor are all variables considered (goodness-of-fit = 66%). This puts it in line with the other Top 3 models, all of which have a higher predictive accuracy than their counterparts. Another clear trend can be seen in imputation: all non-imputed models are superior. Weighting plays only a subordinate role.

The test data sets also confirm that non-imputed data sets are consistently better and that weighting does not play a major role. However, the number of variables does not follow any trend. The best model here is fully weighted, not imputed, and contains all variables (goodness-of-fit = 61%). However, quality of prediction depends above all on imputation.

Table 5.40 Goodness-of-fit (adjusted R^2) best neural networks for NIIS and NIIS Top 3 per weighting and imputation on both full data and test set

Description of parameters varied	NIIS	NIIS Top 3	NIIS (test)	NIIS Top 3 (test)
Fully weighted (Var + Ind +), imputed	61	43	54	36
Fully weighted (Var + Ind +), imputed, individual weighting	50	—	42	—
Indicator weighting (Ind +), imputed	61	38	52	37
Not weighted, imputed	**62**	52	**59**	48
Not weighted, imputed (no revisit, n = 244)	55	52	49	39

Table 5.41 Goodness-of-fit (adjusted R^2) of regressions for IFRI and IFRI Top 3 per weighting and imputation on both full data and test set

Description of parameters varied	IFRI	IFRI Top 3	IFRI (test)	IFRI Top 3 (test)
Fully weighted (Var + Ind +), imputed	22	24	6	17
Fully weighted (Var + Ind +), not imputed	42	53	25	1
Fully weighted (Var + Ind +), imputed, individual weighting	23	—	9	—
Fully weighted (Var + Ind +), not imputed, individual weighting	51	—	**61**	—
Indicator weighting (Ind +), imputed	22	24	6	24
Indicator weighting (Ind +), not imputed	42	51	25	16
Not weighted, imputed	18	20	9	5
Not weighted, not imputed	40	**66**	26	28

5.6.3.2 Random forests

Table 5.42 shows the results of model performance for random forests for the full IFRI data sets.

For random forests, the best model is indicator-weighted, imputed, and only includes the three most important variables (28%). Otherwise, the results are relatively similar to those of other data sets (CPR and NIIS) and there is no variant with an unambiguous influence. On the test data sets, however, model performance varies more strongly. The best model here is fully weighted, imputed, and contains only the three most important variables per factor (goodness-of-fit = 36%).

5.6.3.3 Neural networks

Table 5.43 presents the results for the neural networks models.

The best models, but also the worst, can be found for the Top 3 data. The indicator-weighted, imputed variant achieves the highest predictive accuracy with 31% explanation of variance. On the test data, no robust model seems to exist that performs well. The best model is here indicator-weighted, imputed, and contains only the most important three variables per factor (24% explanation of variance).

To sum up the sensitivity analyses: it could be demonstrated that the various combinations paint a much clearer picture of predictive accuracy. In this way, individual model qualities can be classified and evaluated much better: Is it a robust result or an outlier? Are content-related or methodological reasons the cause for a certain model quality?

For example, although the CPR data show slight variations, they are fairly robust. Model quality ranges from 40% to 60% of variance with a few outliers. The NIIS data is even more robust against all variations and method choices.

Table 5.42 Goodness-of-fit (adjusted R^2) of random forests for IFRI and IFRI Top 3 per weighting and imputation on both full data and test set

Description of parameters varied	IFRI	IFRI Top 3	IFRI (test)	IFRI Top 3 (test)
Fully weighted (Var + Ind +), imputed	20	25	10	**36**
Fully weighted (Var + Ind +), imputed, individual weighting	23	—	14	—
Indicator weighting (Ind +), imputed	20	**28**	11	29
Not weighted, imputed	19	26	23	17

Table 5.43 Goodness-of-fit (adjusted R^2) best neural networks for IFRI and IFRI Top 3 per weighting and imputation on both full data and test set

Description of parameters varied	IFRI	IFRI Top 3	IFRI (test)	IFRI Top 3 (test)
Fully weighted (Var + Ind +), imputed	16	28	10	20
Fully weighted (Var + Ind +), imputed, individual weighting	21	—	15	—
Indicator weighting (Ind +), imputed	17	**31**	15	**24**
Not weighted, imputed	25	7	17	1

The model quality also fluctuates between 40% and 60% of the variance. With the IFRI data, on the other hand, a completely different picture emerges: the results fluctuate very strongly and imputation generally has a very negative effect on model quality.

Based on these variants, it can be concluded that for all sectors (fisheries, forestry, and irrigation), existing success factors can account for about 40%–60% of the variance, sometimes more. Such a predictive accuracy for SES case studies has not yet been achieved for such a large number of cases. However, a lot of improvement is possible—more on this in the next section (Chapter 6). Therefore, it is now possible to continue working on the basis of the models described by picking only those that fulfill certain requirements in terms of methods, robustness, imputation, weighting, or number of variables. On the other hand, using those models as stepping stones, further research could elucidate meta-factors, extend the scope, or use other complementary methods.

6

Discussion and Conclusion

The last section of this book is dedicated to an overall assessment and discussion of the importance of individual success factors. They will also be discussed with regard to the success factor syntheses discussed in the literature in Section 5.1.5. We will present those success factors which proved to be influential and robust over several data sets. The last three sections present the new findings (Section 6.2), an overall summary (Section 6.3), and finally an outlook (Section 6.4) to further research.

6.1 Final assessment

We will start with the overall model itself, where good *participation* possibilities (F13), *dependency on resource* (F10), and easy *manageability* (F5) across all three methods are important. Thus, three out of four factors have been consistently identified as relevant. They are complemented by a *good initial condition of the resource* (F4), which also appears again and again in the top group of important success factors across many data sets and models. Especially for forests, it is of outstanding importance. In addition, *good relationships with others* (F23) and *clear resource limits* (F2) are very relevant. However, they do not reach a top position in all models across all sectors as consistently as the other four success factors.

In practice, systems are often sorted and treated according to their *size* (F1 and F7), their *boundaries* (F2), and their relationships to other groups (F23). Especially with irrigation systems, it is often assumed that size plays an important role in coordination (Lam 1998). In contrast, the present analysis concludes—and this is where all three methods agree—that the size of the system and the number of individuals practically play no role for ecological success (cf. Frey, Villamayor-Tomas, and Theesfeld 2016). This conclusion can also be drawn for the CPR data set. The exception is forest management. Here, the size of the system seems to be important. The most important factor for fisheries is that the limits of a resource system are clearly visible. For forests, the methods do not show any agreement, which suggests that they are of lesser importance. In irrigation systems, on the other hand, clear boundaries are not so relevant.

Sustainable Governance of Natural Resources. Ulrich Frey, Oxford University Press (2020). © Oxford University Press.
DOI: 10.1093/oso/9780197502211.001.0001.

Finally, relations with other actors have been suggested as important: since each resource system with its actor groups is embedded in a social network, it has been assumed that this is a decisive factor (Gruber 2008). The present analysis cannot confirm this for CPR and IFRI—only a medium importance is to be noted. The situation is different for irrigation systems: support from the government, non-governmental organizations, and neighboring groups seems to play a decisive role in their success.

The next paragraphs shortly discuss these most relevant factors for ecological success in light of their importance, even if it already has been discussed why these factors are important (Section 5.1.5).

Real *participation possibilities* ensure that rules can be developed in a precise, fair, flexible, and cost-saving manner. This in turn diminishes the chances that individuals—whether appointed by the government or simply rich individuals—make decisions that are not optimal for the majority of users. However, the importance of this factor could also have been caused by an unintentional unilateral case selection, since the data basis comes from a school of thought that assigns outstanding importance to this factor. One argument against this would be that in all data sets this factor is normally distributed.

In contrast to the relevance of participation in the literature, a strong dependency on resource is discussed controversially—some studies even conclude that strong dependency on resource, in combination with high poverty, promotes excessive exploitation (Agrawal 2007). The models of this analysis, on the other hand, clearly show that a higher dependency leads to handling the resource more carefully. Individuals invest time and effort in it. I suspect that a strong dependency influences many other factors. Furthermore, this factor strengthens self-organization, since individual risks such as bad harvests can be buffered via a strong community (Wade 1992).

The third factor, *manageability*, mainly operationalizes concepts that deal with the costs of withdrawing resource units. This starts with the predictability of when and where the harvest is worthwhile (fish) and when water is expected (irrigation systems). But it is also important how difficult and expensive it is to transport, store, and finally sell or consume the resource units. These more or less favorable spatial and temporal requirements of the units in turn also affect the place of users' residence, monitoring, and resource boundaries, but above all the costs of removal. According to the models calculated here, this is an important success factor.

These success factors are completed by a *good initial condition of the resource* (F4), which is also very relevant across many data sets and models. Especially for forests, it is of outstanding importance. The connection between ecological success and *initial ecological success* is immediately obvious. With already degraded resources (especially forests), it is extremely difficult to prevent individual selfish strategies of exploitation, to set self-organization in motion, and to establish a sustainable, long-term perspective for the future. When users see a resource about to collapse, they tend to exploit as many of the remaining resource units as fast as possible in a race against time and other users (Ostrom, Gardner, and Walker 1994).

In addition, there is the fundamental difficulty of accurately measuring the condition of complex ecological systems. Even for users who have been managing a resource for many years and are deeply familiar with it, they cannot automatically identify trends of deterioration or improvement. In the case of forests, however, it has been shown that a forest expert's assessment corresponds quite well with an elaborate quantitative measurement of the forest's condition (Salk, Frey, and Rusch 2014).

Another identified success factor is an existing *network with other groups and actors* (F23). This ranges from good relations with neighboring groups, thus minimizing conflicts, to a good flow of information to similar users in the region, enabling a group to adopt new techniques or solutions to problems, and good relations with NGOs and the national government. The main advantages are financial support, help by experts, or transfer of know-how.

Considering the classification of success factors once more (see Section 5.1.3, minimum synthesis), it is interesting to see that *one factor of each main category* (resource system, resource units, actors, control system, external environment) is always particularly important. This underlines once more the manifold interactions of many subsystems. It also justifies the selected approach—to analyze these systems as *social-ecological systems*, since both biological and social factors have been considered, even if the biophysical aspects tend to be a bit neglected in comparison to social ones (Epstein et al. 2013).

Now that we have evaluated all success factors in thousands of quantitative models, we would, of course, like to know whether one synthesis, as discussed in Section 2.5, did indeed come up with these five most important success factors? With the exception of the SES framework which lays the theoretical foundations of these databases, *none of the syntheses is able to capture all important factors.* What is interesting is that good opportunities for participation are mentioned in all syntheses and good relations with other actors in almost all. In contrast, the two important success factors affecting biophysical peculiarities of the system (F4, *initial state of the resource* and F5, *manageability*) are hardly represented at all. Finally, dependency on resource is also not mentioned more than three out of ten times.

In conclusion, this means that according to this analysis, many studies place unimportant factors on equal footing with important factors without being able to differentiate between their actual relevance. One reason is that their significance is not derived quantitatively, but usually only from theoretical preliminary considerations or a few case studies, underscoring the need to use quantitative modeling on large-N case study data (Poteete, Janssen, and Ostrom 2010). Table 6.1 presents the importance of success factors across data sets as a summary.

It is, of course, imperative to compare our results with Elinor Ostrom's very well-known *design principles* (Ostrom 1990; Cox, Arnold, and Villamayor Tomas 2010), not only because they figured as blueprints for many studies.

Table 6.1 Ranking of success factors across data sets

Success Factor	CPR	NIIS	IFRI	Combined
F1 Resource size	19	20	2	19
F2 Resource boundaries	2	23	7	3
F3 Accessibility	18	12	8	23
F4 Ecological success at the beginning	4	24	1	8
F5 Manageability	16	3	11	2
F6 Regeneration of RU	5	7	4	15
F7 Number of actors	23	16	5	11
F8 Group composition	3	13	15	6
F9 Social capital	21	4	9	14
F10 Dependency on resource	13	2	12	4
F11 Dependency on group	12	14	23	9
F12 Group boundaries	6	15	20	21
F13 Participation	9	1	3	1
F14 Legal certainty and legitimacy	1	8	16	10
F15 Administration	24	19	13	16
F16 Information	20	18	21	20
F17 Characteristics of rules	15	5	22	7
F18 Fairness	17	6	18	13
F19 Control	14	17	19	24
F20 Compliance	10	10	6	17
F21 Conflict management	22	22	24	22
F22 Exclusion	11	21	10	18
F23 Relations	7	9	14	5
F24 Capabilities to adapt to change	8	11	17	12

Here, too, it is striking that although participation and the social network of the actor group are mentioned a lack of biophysical attributes is also noticeable. In addition, the troika monitoring, conflict management, and gradual punishment are mentioned as important. Surprisingly, all these factors consistently end up in middle and last places in all models considered here. At best, *rule compliance* (F20) could be ascribed medium importance, while *conflict management* (F21) is often found among the last three ranks in the models. This can also be observed in the expert survey on the relevance of factors (Section 4.2): While the experts consider *dependency on resource* (F10) to be important (third place out of twenty-four), which is in line with the models calculated here, factors that have quantitatively been found to be very important like *participation* (F13), *relationship to others* (F23), *manageability* (F5), and *initial state* (F4) are less important in the experts' opinion and end up on the respective ninth, fourteenth, nineteenth, and last position.

Finally, it is of great practical interest to know how a high level of compliance can be achieved. Surprisingly, there are few success factors consistently correlated with it—*participation* (F13), *legal certainty* (F14), and *locally adapted rules* (F17) are among them. The often theoretically predicted close coupling between monitoring and compliance is not clearly present, since only in forests (IFRI) and irrigation systems (NIIS) are there moderate correlations ($r = 0.19$ and $r = 0.18$). In the CPR data set, there is no significant correlation at all.

Across all three data sets—the combined model—only three other success factors are somewhat more strongly correlated with compliance: *participation possibilities* (F13, $r = 0.33$), a *good information flow* (F16, $r = 0.27$), and *fair rules* (F18, $r = 0.27$). Thus, above all, a high level of participation in decisions and good communication between participants, who can give themselves suitable rules, ensures few rule breaks. This causal network has already been suspected and verified in individual case studies (Ostrom 1992a).

Limitations
Like any analysis, this analysis has its limitations.

First, success factors can be arranged in many other structures—there is no right or wrong, but the goal of a comprehensive theoretical framework that is easy to implement may have been missed. Others have to decide on its benefits.

Second, there are also limitations with regard to the data: although the data sets used are probably of a very high quality, that is, they are complete and large data sets, they also contain gaps: CPR lacks about 3% of the data and some IFRI variables contain absurd values despite multiple checks, and some text fields are cut off in NIIS. Except for IFRI, data are also relatively old, which means that it may no longer be fully applicable in today's cases. Imputation on IFRI has shown to decrease model performance considerably.

Third, the operationalization of concepts is always a subjective undertaking due to the large number of individual steps. It is thus possible to arrive at a different assignment of variables to different indicators and subsequently success factors—although hopefully only with small deviations. In order to deal with, for example, subjective weightings etc., measures have been taken at each critical step to make recoding as objective as possible (see Section 3.5.3). Furthermore, for each critical decision in altering the data sets, an alternative was specified and the respective models without the altered data have been calculated. To ensure reproducibility, the process itself has been made as transparent as possible.

Fourth, there are methodological restrictions concerning neural networks. While regressions and random forests represent robust methods, the extraction of the importance of success factors in neural networks is only possible with a high degree of uncertainty.

Fifth, model qualities vary greatly depending on method and data set variant. These differences demonstrate convincingly that uniform results are not possible. In contrast, most results do lie in a certain corridor (goodness-of-fit = 40%–60%). Hence, at least the latter result can be described as robust. It is beyond the scope of this analysis to determine the reasons for these differences.

Sixth, success factors do not include the wider political, economic, and social framework on a national level and long-term processes that go on for decades. This also applies to longer processes in ecosystems—here available data exist for relatively short periods of time, while larger changes point beyond these (Birkhofer et al. 2015). This is, as mentioned, a possible explanation for the model quality for IFRI and the importance of the initial ecological forest condition for success.

6.2 New findings

This section tries to sum up the new findings of this analysis in a short and accessible way and answer the question: Which new insights have been gained?

1. By going beyond existing purely economic review articles (Ledyard 1995; Chaudhuri 2011), a comprehensive synthesis of fundamental factors influencing the willingness of people to cooperate on public goods in laboratories was developed.

There are some overview studies on factors influencing cooperation in PGG. However, they are characterized by a narrow economic perspective and neglect other factors. Therefore, a synthesis of influences such as presented in Section 2.2 has not been available yet.

2. The present analysis expands the research with regard to "design principles" (success factors) in two respects. First, a theoretical synthesis of mechanisms of action (5.1.5) clarifies how certain factors have an influence on ecological success in the management of natural resources. Second, with this synthesis it could be derived in the future how influencing factors interact in causal structures.

Although Elinor Ostrom's concept of design principles is one of the most influential within research on SES (Ostrom 1990) and was decisive for the award of the Nobel Prize, the number of follow-up studies have remained limited. Thus, although there is a meta-study that examines the effectiveness of these principles and arrives at consistently positive results (Cox, Arnold, and Villamayor Tomas 2010), and although there are also various individual studies that use her design principles (an expression that according to Elinor Ostrom she would have preferred to have withdrawn, which is one reason why the present analysis uses the term success factors instead) as structuring measures (e.g., Nilsson 2001), a systematic analysis of success factors on a theoretical level does not exist, as presented here. Many of the meta-analyses mentioned in this paper (e.g., Brooks, Waylen, and Borgerhoff Mulder 2012; Cinner et al. 2012; Gutiérrez, Hilborn, and Defeo 2011) are also based on Ostrom's previous work and use parts of the SES framework. However, I am not aware of any study that attempts a systematic and theoretically sound derivation of factors as found in Section 5.1.5 and Table 5.1, although the studies from Pagdee, Kim, and Daugherty (2006) and Baland and Platteau (1996) are somewhat similar.

3. This study presents one of the few syntheses of success factors that has been *empirically* tested.

Those few syntheses of success factors that exist are discussed in Section 2.5, but neither of them is, for the purpose of this analysis, comprehensive enough. The present analysis offers a synthesis of success factors in which practically all concepts mentioned in this literature can be placed on one of the different tiers. Attention is paid to the degree of abstraction, scope, and theoretical significance of concepts (Section 4.2.3). This results in a comprehensive, balanced synthesis of influencing factors that can be used, for example, as data structures for field studies. In addition, these twenty-four factors are not only extracted from the literature but also subjected to extensive empirical tests (see Sections 5.2, 5.3, 5.4, 5.5, and 5.6).

4. A major research gap is the lack of operationalization of many factors in SES. For the first time, this paper creates a uniform and graduated indicator system for a high number of potential success factors.

In previous works, many concepts (such as social capital, dependency on resource, etc.) have been operationalized very differently or even not at all. This analysis presents a comprehensive indicator system (Section 4.2) that is validated through expert surveys (e.g., by Elinor Ostrom). In addition, role and location of each factor within this indicator-based conceptual system is specified by precise definitions and elucidation of their relationships (Section 5.1.5). This guarantees the reproducibility of results, and for the first time a comprehensive indicator system for local SES is made available.

5. One of the main objectives of this analysis—a general, quantitative, precise, and robust model for natural resource management—has been achieved. The models achieve high to very high explanatory quality for all three data sets. In addition, by creating a cross-sectoral overall model for the first time, various hypotheses and research questions can be answered.

One of the most striking research gaps in the field of SES is the lack of general quantitative models that are based on a broad database and have satisfying explanatory quality. One reason for this is that individual case studies predominate and very few large data sets are available. Furthermore, until now no methodology existed to model unknown, non-linear inter-factor interactions. Nor was an operationalized synthesis of success factors available. The present analysis solves each of these individual problems and explains variances of 81% for CPR, 62% for NIIS, 66% for IFRI, and 46% for the overall model.

Unfortunately, a comparison of model qualities to other studies is hardly possible, since previous experiments only consider a much smaller set of influencing factors and thus only get a partial view of successes (e.g., van Laerhoven 2010; Chhatre and Agrawal 2008). Others do not provide a model despite a synthesis of success factors (Pagdee, Kim, and Daugherty 2006). The only existing comparison is a meta-analysis of co-managed fisheries published in *Nature*, which—also with random forests—can explain 71% of the variance in 130 case studies (Gutiérrez, Hilborn, and Defeo 2011).

6. This analysis enables the classification of SES with high accuracy in regard to their ecological success.

The separation into training and test data sets also allows to classify whether a system is unsuccessful, average, or successful. This can be helpful for many measures in case studies. Here, a correct classification for CPR is achieved for 88% of test cases, 58% of NIIS, and 77% of IFRI. This is important for many practical applications, such as policy measures according to potential future developments.

7. The models created can be used as an easy-to-use policy advice tool. This makes it possible to assess the effects of certain measures or interventions in a system (e.g., investments or legal changes) before they are actually implemented.

The models of trained neural networks allow the manipulation of each individual success factor for each individual case. Thus, the models are ready to be used by laymen and allow them to test various scenarios: Does an improvement of the infrastructure bring about a significant increase in ecological success? Is social capital crucial? Which factors generally offer the greatest potential for improvement? The calculated models are able to give precise answers (within approximately 6% uncertainty) individually for each system.

By using existing graphical user interfaces (e.g., in Membrain), each factor of the system of interest can be assigned a value between -1 (very negative) and 1 (very positive). This directly results in the estimated ecological success by the trained model. Thus—before expensive or lengthy measures are taken in real life—changes in success factors may be evaluated in regard to their effect on the overall system. Not only may single success factors be changed, but so, too, can groups of factors. This is particularly important, since changing certain combinations sometimes shows surprising results due to their non-linear interactions.

This could make many independent public good projects, projects promoted by NGOs, or development cooperation more successful since the most relevant factors for a particular project could be determined in advance. This clarification of local success factors should be of interest for the institutions themselves, but also for external actors like the national government. If a structural improvement is to be implemented on a large scale, estimating its effects before investing millions of dollars could be worthwhile. Thus, this analysis could contribute to an improved sustainability and efficiency of existing and future public domain projects.

8. The present analysis suggests a need to re-evaluate the relevance of some factors which are considered to be very important. These include monitoring or conflict resolution mechanisms.

Some success factors are attributed outstanding importance in the literature, such as monitoring (Chhatre and Agrawal 2008). In one meta-analysis, monitoring is also the factor with the largest effect size (Cox, Arnold, and Villamayor Tomas 2010). However, in another methodologically very clean meta-analysis (Gutiérrez, Hilborn, and Defeo 2011), monitoring occupies only the seventh of twelve places in terms of relevance for success in 130 fisheries. The present work, which is based on an even broader data basis, confirms this rather subordinate importance of monitoring in contrast to the majority of the studies on this factor. In no model is monitoring of decisive importance—in the ranking by random forests, this factor ranks seventeenth for CPR and NIIS, nineteenth for IFRI, and even last in the overall model (Tables 5.3, 5.11, 5.19, and 5.27). This small relevance also applies to conflict resolution mechanisms. The complex reasons for this are outside the scope of this analysis—it may be that a positive effect may cancel out negative cases, it might be connected to size, etc.

In general, since some factors have been identified as decisive, others as rather unimportant, future research is now able to concentrate on important factors and to explain better their causal interplay. Future data collection could also benefit by focusing on these very important factors. This in turn would enable more precise indicators of success.

9. Theories regarding SES have been expanded. A detailed and comprehensive synthesis that can be directly operationalized is now available (see Section 4.2.3).

The synthesis of success factors may guide data collection or fulfill other case study requirements. As such it has already been used in research projects, for example in England (Aglionby 2014) and Africa. It might also help to aid ongoing work of the widely used SES framework.

6.3 Summary

Despite a wealth of individual results, the field of SES is still not developed in some respects. Prominent unsolved problems are the lack of comparability between case studies, the difficulty to deal with interactions between influencing factors in SES, and the open question of which factors play a role at all. This paper addresses these intertwined problems and tries to answer the question which factors are relevant for ecological success in the management of natural resources. This should ultimately prevent loss of habitat and biodiversity with regard to ecosystems (Sala et al. 2005), while at the same time ensuring sustainable extraction and subsistence management.

In a first step, we investigated fundamental biological influences on cooperative behavior. The focus here was on experimental social dilemma situations, which, from a game-theoretical perspective, underlie real decision-making situations in SES. This synthesis on an individual level already draws attention to a number of potential success factors that also play an important role at the group level, such as reputation and communication. These point towards the importance of the social group structure, which in turn has a major impact on the rule system.

Our next step was to give an overview of previous research on success factors. There are a few clear-cut results. First, studies do not necessarily agree on a minimum set of influencing factors between studies. Second, there is no common operationalization of factors. Third, the empirical basis of many theoretical statements is thin. Fourth, many studies are limited to a few variables. The present analysis solves these problems through a very broad and uniform basis of case studies (n = 794), the use of many variables, and a uniform operationalization of concepts.

We chose three high-quality empirical data sets as basis for statements on generalizable patterns in case studies. In a lengthy process, we checked data for correctness, then selected and recoded data, and then put this recoded data through various procedures—such as different weightings, the use of the three most important variables only, replacement of missing values, and finally splits in training and test data sets. This ensures that no single methodical step could distort our results. This multitude of data sets, provided by many different automated methods, was then analyzed with different statistical methods.

We used three methods of analysis: multivariate linear regressions, random forests, and artificial neural networks with regressions being mainly used as reference and benchmark for the other two methods. We chose random forests because they represent very robust methods that can handle missing values well and allow direct estimates of any individual factor's relevance. We used our third method, neural networks, because they perform well whenever it is unclear which interactions between possible influencing factors exist—particularly if they are presumably non-linear in nature. Moreover, they often provide the best model quality for machine learning algorithms. By using several methods, we ensured that all model results estimates are more robust since the advantages of the individual methods compensated for the others' disadvantages. This concerns, for example, the linearity of regressions or the problem of factor extraction in neural networks. Figure 6.1 shows goodness-of-fit across data sets and methods.

This paper expands the field of analysis of SES by presenting a synthesis of system attributes in the sense of Ostrom's design principles (Ostrom 1990), which are called success factors and are of potentially high relevance for the success of SES. As with Ostrom, the focus is neither on self-organization, nor social and economic success, but on ecological success. The creation of such a success

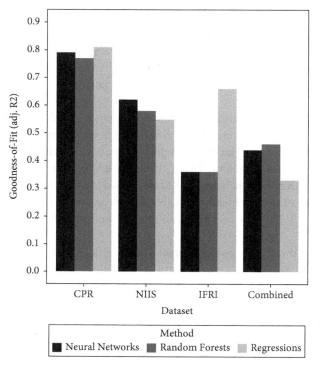

Figure 6.1 Model quality across data sets and methods

factor synthesis and its validation via empirical case studies has been attempted several times (Agrawal and Chhatre 2006; Cinner et al. 2012; Gutiérrez, Hilborn, and Defeo 2011; Brooks, Waylen, and Borgerhoff Mulder 2012). Although there is no sign of convergence, as various factors are judged to be differently relevant, some factors are now judged to be important with a certain regularity in various studies. These are, above all, good participation possibilities, a high dependency on the resource and easy manageability of resource units. This analysis hopes to contribute to these studies and, above all, to go beyond them in methodological terms, since a very broad and robust picture has emerged through the use of many methodical precautions and variants during model development.

Overall, this analysis can prove twenty-four success factors as relevant concepts for ecological success in SES. This is demonstrated by the models' high goodness-of-fit. The results also show that there is no single factor that is extremely important in comparison to all others. This also applies to a combination of factors—although this analysis filters out four to six factors having overarching importance. However, as the results for IFRI show, these are not a panacea for success either. Hence, these results also answer the question of whether there are models that can adequately describe complex SES regardless of sector (fishing, irrigation, and forest management): success factors do exist on a general level. It can therefore be abstracted from individual systems; at the same time, however, it becomes clear that there are limits to this generalization. Sector-specific models with less data are indeed better than a general model.

In terms of model development, this analysis goes beyond previous studies. First, it considers a more comprehensive range of factors, providing a synthesis of most of the attributes discussed in the literature. Many variables are assigned to abstract concepts—the success factors—and are operationalized. This also makes it easier to assess the importance of factors in other studies.

Second, in contrast to existing studies, the number of variables (several hundred) used is very high, providing a much more detailed picture of SES. At the same time, models have been calculated with the three most important variables only. It turns out that in some cases the detailed representation of concepts is better, in some cases worse. Thus, for the first time, the complexity of SES can be captured using models containing hundreds of variables within a comprehensive set of potential success factors. Going beyond a few case studies, each of which identifying a few variables as relevant, is a step that has often been called for (Poteete, Janssen, and Ostrom 2010).

Third, this analysis uses a high number of case studies to reach its conclusions. This very broad, empirical basis of high-quality case studies allows analysts to assess the results as reliable.

Methodologically, the attempt to capture the complexity of such systems in non-linear models and to describe them in general form on an abstract level has been successful. The present analysis aimed at this generalization. Especially neural networks have proven to be consistently superior to conventionally used methods such as linear regressions, with sometimes very large differences in model quality. Together with the fact that it has been possible to extract the significance of the factors from these black box models, this analysis has added another valuable tool to the SES analysis toolbox.

At the beginning of this book we formulated three major working hypotheses. All three working hypotheses have been fully confirmed. Our first hypothesis asked whether general patterns in SES did indeed exist. We could establish that acceptable models exist across many case studies. Since the models explain up to 81% of the variance (neural networks for CPR), and very good models also exist for NIIS and IFRI data with 62% and 66% explanation of variance, it is obvious that there are indeed success patterns across cases. Hence, we can conclude that a retreat to purely case-specific influences, which are presented as not generalizable, is no longer tenable (as stated, e.g., by Cleaver 2000).

The second hypothesis aimed at whether the postulated twenty-four success factors are suitable to explain ecological success in SES. Here, too, the successful models show that there is a close relationship between potential factors investigated for ecological success and the actual condition of the respective resource—be they forests, fisheries, or irrigation systems. Especially four factors (see previous argument) have been demonstrated to be important.

Finally, the third hypothesis postulated non-linear relationships within and between success factors, that is, a complex network of relationships. If that were the case, methods such as neural networks should therefore be superior to conventional statistical methods in their predictive accuracy. This hypothesis was also fully confirmed. Model quality of neural networks was sometimes clearly higher than that of other methods, although some single configurations of other methods (see Figure 6.1) show very high explanatory power.

6.4 Outlook

In this last section of the book, I would like to point out some directions of research that follow naturally from the preceding analyses or that are important in my opinion for advancing SES research.

Obviously, a detailed analysis of success factors that have been identified as very important (*fairness* and *legal certainty* for fisheries and irrigation systems, *participation* and *resource dependency* for irrigation systems in Nepal, and *initial state of the forest* and *number of actors* for forest management) should prove to be particularly fruitful. If they were investigated and improved, systems could profit directly.

These combinations of factors alone point to very different ways of intervention in some cases: the primary objective in fisheries might therefore be to establish a fair and externally recognized system of rules. In Nepalese irrigation systems, on the other hand, it is more a matter of promoting commitment (e.g., through water user associations), and in forests, the biological conditions for forest growth seem to be decisive, besides the number of actors.

The synthesis of our twenty-four success factors could also be used in future case studies and contribute to improved data collection and modeling. On the one hand, a data set of empirical studies could be generated which then could be coded and analyzed on the basis of these success factors. Given the small number of existing meta-analyses such a study would be of great interest, not only because it would represent a further, independent examination of the generalizability of the twenty-four success factors. On the other hand, with our analysis, it is possible to structure and implement new field studies and data surveys. If research were to use our results, the general SES basis of data could be expanded by enabling comparative case analyses.

Many models, including the models presented here, do not account for the larger frame outside the local SES. A substantial part of the unexplained variance can be explained by regional or national socio-economic surroundings. If SES-researchers were to include framework conditions at a higher level and add economic or political data—national indicators such as gross domestic product (GDP), Human Development Index (HDI), Corruption Perceptions Index (CPI)—to local success factors, we could sharpen our perception of what is happening in and around SES (see, e.g., Gutiérrez, Hilborn, and Defeo 2011).

Moreover, modeling could not only be extended to include framework conditions at a higher level but could also be narrowed down to individual factors. An intermediary step would be analyses performed at the household level, but more research on decision-making at the individual level is also needed. For example, there are numerous personality traits in psychology (such as the five-factor model, "big five") that may explain cooperative behavior on an individual basis. Coupling local-level SES research with individual decision-making traits would probably also contribute to a clearer specification of models concerning scope and level of analysis.

There are two further research gaps in SES research that need to be addressed before a deep understanding of SES can be reached. On the one hand, there is little research on the causal pathways between success factors. Here, the success factors and their effects in detail (see Section 5.1) could be a first step. Be it through studies of their exact interdependencies, through case studies with special consideration of these factors, or by identifying those system configurations in which certain success factors play a special role—they could be the basis for a first analysis of causal paths based, for example, on the correlations between factors. One modeling technique able to do that is known as structural equation modeling (SEM). Such approaches in turn would shed light on the too static success factors.

Second, there is far too little *formalization* of SES. Formalization would improve cooperation within the research community as content would be available in structured, uniform, and reusable form. This may be achieved by formalizing its entities, relationships, and attributes which would result in a computer-based ontology (Frey and Cox 2015). In turn, this requires a logically consistent definition of concepts and their relationships (Hinkel et al. 2015). Once this is achieved, automated algorithms can search for errors and automatically create new relationships that are based on existing rules. Content could then be analyzed independent of language. This could lead to better cooperation within the research community, as content is available that is structured, uniform, and reusable.

Finally, in recent years the SES research community has demanded the development of a common base of data in form of an easily usable database. The theoretical preparatory work for such an implementation shows that a common conceptual basis is extremely relevant. The synthesis presented here could also be used profitably for this purpose, next to, for example, the SES framework. The CPR, NIIS, and IFRI databases have shown that data collection has to be informed by a sound theoretical framework, and they have also demonstrated that longitudinal data is of great importance. At the moment, there are efforts to gather revisits-data for Nepal and to build up a database for Asian irrigation systems, like the NIIS. There are also overviews about existing databases (Partelow 2018).

And last but not least, neural networks have proven to be superior to conventional methods in many cases. In this sense, it is to be hoped that their use—and machine learning methods in general—will gain traction in the community. With machine learning as new tool, we can hope that more precise SES models are possible in the near future.

7

Appendix

The Appendix is structured in the same way as the main text. References from the introduction can therefore also be found in the Appendix under the introduction, that is Chapter 1.

3 Data

For better readability, some technical transformations of the data have been removed from the main text. They are listed here in the Appendix.

3.1 Recoding of variables—Transformation

- *Yes/No questions*: are coded using an Excel formula with −1 and 1, depending on the question and relation to the success factor. Example: "In your estimation are the rules-in-use: easy to understand by the appropriators? This variable is assigned success factor 17 (Adapted rule system). Yes is coded with 1, and no is coded with −1.
- *Matrices*: Database fields that are in matrix or group notation are separated in a separate tab by Excel formulas and then recoded. Example: "In the following list mark which activities the members of this organization (or group) use to express their needs and concerns to those officials of this organization who make collective choice decisions in relation to the resource. Use the code outlined above (question A7)." Possible answers are: "Elections; Formal petitions; Formal hearings; Advice and consent on nominations to nonelected positions; Demonstrations; General meetings; Illegal exchanges with officials; Informal contacts; Other." The corresponding database field has the format "30000003.4." Each digit of this matrix indicates a possible response. The possible values are: " '0': that activity is not used. '1': that activity is used and appropriators have a favorable evaluation of its usefulness. '2': that activity is used and appropriators have an unfavorable evaluation of its usefulness. '3': that activity is used and the document does not indicate the appropriators' evaluation of its usefulness."

Sustainable Governance of Natural Resources. Ulrich Frey, Oxford University Press (2020). © Oxford University Press.
DOI: 10.1093/oso/9780197502211.001.0001.

Here again, each individual field/value pair is encoded differently using an Excel formula.

- *Likert scales*: are recoded with an Excel formula, which are classified on the scale from −1 to 1 depending on the evaluation of the answers for the success factor. Example: "How frequently do the functionaries of the main and branch committee meet with the users to identify their problem?" The possible answers are here: "Very rarely (1); At least once a year (2); At least once in each irrigation season (3); Several times in each irrigation season (4)." The more frequent the meetings, the more positive this is for the opportunities for participation.

- *Numbers*: In the case of absolute numerical values, such as resource size or the number of actors, there is no normal distribution; instead, there are some very large systems or very many actors. Normalization leads to many small systems being assigned small values close to −1 and being hardly distinguishable. For this reason, each variable with absolute numbers is first sorted, then visualized as a scatterplot and thus determined in which area most systems are located or where the limit for the few large systems is (natural gap in the data). This limit is then defined as the new upper limit; all larger systems are assigned the maximum achievable value (e.g., 1). This was possible for all variables without any problems—there was always an obvious natural gap. This equalization leads to a better differentiation of the individual cases via the data values. This only takes place for two variables (size of the system, number of users). In all other cases, the existing numeric characteristic values are simply transformed using a normalization function $(−1+2*ABS(Value − (Min))/ABS(Max − (Min)))$.

- *Text*: The recoding of the text fields is described in the main text, Section 3.5.3.2.

- *Likert scales 2*: By far the most common case of variable recoding is the special handling of Likert scales that are not strictly arranged in ascending or descending order. An example is RESOURCE ANALUNIT ("What is the analytical type of unit?"): The four possible response values are (1) Renewable stationary, (2) Renewable moving (fugitive), (3) Non-renewable stationary, and (4) Non-renewable moving (fugitive). For the success factor manageability with the criterion ease of harvesting and the indicator mobility, the distinction between stationary (answers 1 and 3) and volatile (answers 2 and 4) is important. Obviously, (1) and (3) are cheaper than (2) and (4) for a simpler harvest. They are recoded in Excel using standardized formulas (here according to: 1 and 3 becomes 1; 2 and 4 becomes −1).

4 Methods

4.1 Expert survey and expert evaluation

Table 7.1 Mean value of relevance according to the rating of five experts regarding the success factors for ecological success

Success factor	Mean
Resource	
Size of resource system	6.7
Clarity of resource boundaries	8.5
Condition at the beginning (maintenance level, rate of withdrawal, externalities)	5.3
Accessibility (barriers etc.)	6.3
Resource Units	
Characteristics of resource units (visibility, distribution)	7.3
Level of regeneration (time needed for units to be appropriated)	6.7
Manageability (ease of harvesting, accessibility, mobility, storage possibilities)	6.6
User Group	
Number of Users	5.6
Group composition (heterogeneity (cultural, economic, of interests))	5.5
Dependence (on resource, on group)	8.8
Social capital (trust, group cohesion, shared moral norms, long term commitment, common history, degree of networking of users)	9.2
Motivation (benefits, incentives, willingness to invest)	6.9
Rule System	
Clarity of group boundaries	7.2
Participation (existence of institutions, rights, arenas, and locally adapted rules)	8.1

Continued

Table 7.1 *Continued*

Success factor	Mean
Leadership (local, experienced, trusted)	9.1
Information (about resource, resource units, user group, ease and effectiveness of communication)	8.3
Legal certainty (existing rights, level of corruption)	7.6
Characteristics of rules (rights to establish own rules, clarity, simplicity, ease of understanding, flexibility)	6.9
Compliance (rule following, monitoring, enforcement, sanctions, feedback system, level of corruption)	8.3
Fairness (equity, level of trust)	7.5
Conflict management (level of conflict, existence of adjudication, arenas, jurisdiction, mediation from outside, clear rules)	7.5
External Environment	
Exclusion (existence of rights, ease of exclusion, number of infractions)	8.3
Legitimacy (existence of own rights, recognition by state)	8.3
Support/Conflict (collaboration with state, NGOs, other appropriators)	6.9
Adaption to capabilities to change (nature, markets, technology)	6

4.2 Variables for ecological success (F25)

The dependent variable has a special importance. Therefore, for each data set, a detailed description of all variables included in this concept is given—the name of the variable in the database, the type of variable, and a brief description. The description is abbreviated for space reasons, the actual description or question is often longer and provided with additional explanations.

4.2.1 CPR

Table 7.2 CPR—Description of variables in ecological success (F25)

Variable name	Type of data	Short description
loc_ENDDATE	Number	Begin and End date (end)
opl_BEGDATE	Number	Begin and End date (beginning)
Opl_BMARKETS	Likert scale	How are the appropriated units disposed of (beginning)?
opl_CONDITON	Likert scale	Physical condition of the system
opl_EAVERAGE	Number	Average age of the units withdrawn from this resource at the end
opl_EAVERSIZ	Number	Average size of the units withdrawn from this resource at the end
opl_ECONEFF	Likert scale	Short–run Economic Technical Efficiency
opl_effindc	Text	Indicators and means of increasing efficiency
Opl_EMARKETS	Likert scale	How are the appropriated units disposed of (end)?
opl_ENDBLNC	Likert scale	Balance between quantity of units withdrawn and number available (end)
opl_ENDCONDA	Likert scale	How well–maintained is the appropriation resource (end)?
opl_ENDCONDD	Likert scale	How well–maintained is the distribution resource (end)?
opl_ENDCONDP	Likert scale	How well–maintained is the production resource (end)?
opl_ENDDATE	Number	Beginning and ending of the operational level
opl_ENDNTFER	Likert scale	Interference between technology and processes for other resources (end)
opl_ENDPOLL	Likert scale	Problems of pollution (end)
opl_ENDQUAL	Likert scale	Quality of units being withdrawn (end)
opl_ENDRATE1	Number	Volume of withdrawal for fisheries (end)
opl_ENDRATE3	Number	Volume of withdrawal for irrigation (end)
opl_ENDTECHX	Likert scale	Extent of technical externalities (end)
opl_ESEXDEVL	Likert scale	Are the units sexually mature at this size or age (end)?

Continued

Table 7.2 *Continued*

Variable name	Type of data	Short description
opl_Evaluate	Text	Brief synopsis of how this system is evaluated (performance)
opl_MTONHA	Number	Metric tons of agricultural product per year per hectare
opl_NEWTECH	Likert scale	Is new technology introduced?
opl_NEWVALUE	Likert scale	External change in exchange value of units appropriated?
opl_ONEMARKT	Likert scale	Do appropriators sell this unit in more than one market?
opl_TAILEND	Likert scale	Adequacy and predictability of water to tailenders
opl_TECHEFF	Likert scale	Technical Effectiveness of water availability
opl_TYPRESUL	Text	Evaluation of results
res_MULTAPPR	Likert scale	Relationship among multiple appropriation processes
res_WHENBILT	Number	Date of construction of system
sbg_LGTHUSE	Likert scale	Length of time this subgroup has regularly appropriated
scr_paragrph	Text	Abstract of document being screened

4.2.2 NIIS

Table 7.3 NIIS—Description of variables in ecological success (F25)

Variable name	Type of data	Short description
agr_Cr1HYV	Number	% of farmers using HYVs for crop 1
agr_Cr2HYV	Text	% of farmers using HYVs for crop 2
agr_Cr3HYV	Text	% of farmers using HYVs for crop 3
agr_Cr4HYV	Text	% of farmers using HYVs for crop 4
agr_Cr5HYV	Text	% of farmers using HYVs for crop 5
agr_Cr1Max	Number	Maximum Yield (t/ha) crop 1
agr_Cr2Max	Number	Maximum Yield (t/ha) crop 2
agr_Cr3Max	Text	Maximum Yield (t/ha) crop 3
agr_Cr4Max	Text	Maximum Yield (t/ha) crop 4
agr_Cr5Max	Text	Maximum Yield (t/ha) crop 5
agr_Cr1Min	Number	Minimum Yield (t/ha) crop 1
agr_Cr2Min	Number	Minimum Yield (t/ha) crop 2
agr_Cr3Min	Text	Minimum Yield (t/ha) crop 3
agr_Cr4Min	Text	Minimum Yield (t/ha) crop 4
agr_Cr5Min	Text	Minimum Yield (t/ha) crop 5
agr_Cr1Avg	Number	Average Yield (t/ha) crop 1
agr_Cr2Avg	Number	Average Yield (t/ha) crop 2
agr_Cr3Avg	Text	Average Yield (t/ha) crop 3
agr_Cr4Avg	Text	Average Yield (t/ha) crop 4
agr_Cr5Avg	Text	Average Yield (t/ha) crop 5
loc_ENDDATE	Number	Year in which fieldwork was completed
opl_Enddate	Number	Date of field research
opl_ENDqual	Number	Quality of units being withdrawn
opl_ENDpoll	Number	Problems of pollution (end)
opl_ENDconda	Number	How well-maintained is the appropriation resource?
opl_ENDcondD	Number	How well-maintained is the distribution resource
opl_ENDcondP	Number	How well-maintained is the production resource
opl_typeresul	Text	Evaluation of results

Continued

Table 7.3 *Continued*

Variable name	Type of data	Short description
opl_effindc	Text	Indicators and means of increasing efficiency
opl_TechEff	Number	Technical Effectiveness of water availability
opl_EconEff	Number	Short–run Economic Technical Efficiency
opl_Conditon	Number	Physical condition of the system
opl_TailEnd	Number	Adequacy and predictability of water to tailenders
opl_Evaluate	Text	Brief synopsis of how this system is evaluated (performance)
opl_HeadInt	Number	Influence of the Monsoon on irrigation (head)
opl_TailInt	Number	Influence of the Monsoon on irrigation (tail)
opl_shead	Number	Water availability (spring) at the head
opl_sriceh	Number	Water availability for rice (spring) at the head
opl_svegh	Number	Water availability for veg (vegetables) (spring) at the head
opl_smaizeh	Number	Water availability for maize (spring) at the head
opl_soch	Number	Water availability for och (spring) at the head
opl_sfallowh	Number	Water availability for fallow (spring) at the head
opl_stail	Number	Water availability (spring) at the tail
opl_sricet	Number	Water availability for rice (spring) at the tail
opl_svegt	Number	Water availability for veg (vegetables) (spring) at the tail
opl_smaizet	Number	Water availability for maize (spring) at the tail
opl_soct	Number	Water availability for och (spring) at the tail
opl_sfallowt	Number	Water availability for fallow (spring) at the tail
opl_mhead	Number	Water availability (Monsoon) at the head
opl_mriceh	Number	Water availability for rice (Monsoon) at the head
opl_moch	Number	Water availability for och (Monsoon) at the head
opl_mfallowh	Number	Water availability for fallow (Monsoon) at the head
opl_mtail	Number	Water availability (Monsoon) at the tail
opl_mricet	Number	Water availability for rice (Monsoon) at the tail
opl_moct	Number	Water availability for och (Monsoon) at the tail
opl_mfallowt	Number	Water availability for fallow (Monsoon) at the tail

Table 7.3 *Continued*

Variable name	Type of data	Short description
opl_whead	Number	Water availability (winter) at the head
opl_wwheath	Number	Water availability for wheat (winter) at the head
opl_wvegh	Number	Water availability for veg (vegetables) (winter) at the head
opl_wmaizeh	Number	Water availability for maize (winter) at the head
opl_woch	Number	Water availability for och (winter) at the head
opl_wfallowh	Number	Water availability for fallow (winter) at the head
opl_wtail	Number	Water availability (winter) at the tail
opl_wwheatt	Number	Water availability for wheat (winter) at the tail
opl_wvegt	Number	Water availability for veg (vegetables) (winter) at the tail
opl_wmaizet	Number	Water availability for maize (winter) at the tail
opl_woct	Number	Water availability for och (winter) at the tail
opl_wfallowt	Number	Water availability for fallow (winter) at the tail
opl_mtonha	Number	Metric tons of agricultural product produced per year per hectare
opl_winflow	Number	Volume of peak, winter, and low flow of the source (winter)
opl_lowflow	Number	Volume of peak, winter, and low flow of the source (low flow)
opl_peakflow	Number	Volume of peak, winter, and low flow of the source (peak)
ors_Wssstime	Text	Timeliness in irrigation delivery
ors_Wsssuppl	Text	Adequacy of irrigation water supply
ors_Wssrelia	Text	Reliability of irrigation water supply
res_WHENBILT	Text	Date of construction of system
res_PARAGRPH	Text	Abstract of case being coded
sbg_LGTHuse	Number	Length of time this subgroup has regularly appropriated
scr_paragrph	Text	Abstract of document being screened

4.2.3 IFRI

Table 7.4 IFRI—Description of variables in ecological success (F25)

Variable name	Type of data	Short description
FOREST.FTREEDENSY	Likert scale	Has the density of the forest trees changed in the past 5 years
FOREST.FBUSHDENSY	Text	ˈDensity of shrubs and bushes change: top 3 reasons
FOREST.FCOVDENSY	Text	Forest ground cover change: 3 top reasons
FOREST.FPLANTED	Yes/No	Was this forest originally planted?
FOREST.FPROBLEMS	Text	Serious problems for users during next five years?
FOREST.FOPPORTUN	Text	Greatest opportunities for users during next five years?
GRPTOFOR.GPROBLEMS	Text	User group's & management's estimate of serious problems in next 5 years
GRPTOFOR.GPROBLEMS	Text	User group's estimate of serious problems in next 5 years
GRPTOFOR.GOPPORTS	Text	User group's estimate of greatest opportunities in next 5 years
GRPTOFOR.GHISTCHNG	Text	Major changes in relationship between forest and user group
INTERORG.ICONFLUSE	Text	Do conflicts among user groups affect use of forest?
OVERSITE.OSITELAT	Text	Latitude of site
OVERSITE.OSITELONG	Text	Longitude of site
OVERSITE.OSITEELEV	Text	Elevation of site
F_SPECIE.F_BIONAME	Text	Harvested plants disappeared 5 years ago: Botanical Name
F_SPECIE.F_LOCNAME	Text	Harvested plants disappeared 5 years ago: Local Name
F_SPECIE.F_NUMYEARS	Number	Harvested plants disappeared 5 years ago: Number Years
F_SPECIE.F_REASON	Text	Harvested plants disappeared 5 years ago: Reason
F_SPECIE.F_BIONAME	Text	Harvested plants disappeared 10 years ago: Botanical Name
F_SPECIE.F_LOCNAME	Text	Harvested plants disappeared 10 years ago: Local Name

Table 7.4 *Continued*

Variable name	Type of data	Short description
F_SPECIE.F_NUMYEARS	Number	Harvested plants disappeared 10 years ago: Number Years
F_SPECIE.F_REASON	Text	Harvested plants disappeared 10 years ago: Reason
F_SPECIE.F_BIONAME	Text	Harvested plants disappeared 15 years ago: Botanical Name
F_SPECIE.F_LOCNAME	Text	Harvested plants disappeared 15 years ago: Local Name
F_SPECIE.F_NUMYEARS	Number	Harvested plants disappeared 15 years ago: Number Years
F_SPECIE.F_REASON	Text	Harvested plants disappeared 15 years ago: Reason
FOREST.FTREEDENS	Yes/No	Has the density of the forest trees changed in the past 5 years
FOREST.FBUSHDENS	Yes/No	Has the density of the forest bushes changed in the past 5 years
FOREST.FCOVDENS	Yes/No	Has the density of the forest ground cover changed in the past 5 years
FOREST.FVEGCHANGE	Yes/No	Has the area of the forest changed in the past 5 years
FOREST.FINCREASE_	Text	If area increased, what are the reasons?
FOREST.FINCOTH	Text	Other reasons
FOREST.FDECREASE	Text	If area decreased, what are the reasons?
FOREST.FDECOTH	Text	Other reasons
F_ORGAN.F_BIONAME	Text	Master List of Plant Species Botanical Name
F_ORGAN.F_FAMILY	Text	Master List of Plant Species Family Name
F_ORGAN.F_LOCNAME	Text	Master List of Plant Species Local Name
F_ORGAN.F_TYPE	Text	Master List of Plant Species TYPE
F_ORGAN.F_IMPORT	Text	Master List of Plant Species Reason Important
F_ORGAN.F_ABUNDANT	Yes/No	Master List of Plant Species Is Abundant
F_ORGAN.F_USES	Text	Master List of Plant Species Use
F_ORGAN.F_BIONAME	Text	Master List of Animal Species Scientific Name

Continued

Table 7.4 *Continued*

Variable name	Type of data	Short description
F_ORGAN.F_LOCNAME	Text	Master List of Animal Species Local Name
F_ORGAN.F_TYPE	Text	Master List of Animal Species TYPE
F_ORGAN.F_IMPORT	Text	Master List of Animal Species Reason Important
F_ORGAN.F_USES	Text	Master List of Animal Species Use
F_INORG.F_ABUNDANT	Yes/No	What other resources are found in the forest? Is Abundant
FOREST.FVEGDENSE	Likert scale	The density of vegetation in this forest is:
FOREST.FSPECIEDIV	Likert scale	The species diversity in this forest is:
FOREST.FVALUECOM	Likert scale	The commercial value of this forest is:
FOREST.FVALUESUB	Likert scale	The subsistence value of this forest is:
FOREST.FCONSERVE	Likert scale	Type of conservation measures
PLOT.PEROSION	Yes/No	Plot data: Is there active soil erosion?
PLOT.PLIVESTOCK	Yes/No	Plot data: Evidence of livestock use?
PLOT.PINSECTS	Yes/No	Plot data: Evidence of damage by insects/pests?
PLOT.PCONDITION	Text	Plot data: Plot conditions
PLOT.PEPIPHYTES	Likert scale	Plot data: Epiphytes
P_GCOVER.FK_F_ORGAN	Text	Botanical Name
P_GCOVER.P_TYPE	Text	Ground cover: Plant Type
P_GCOVER.P_PERCENT	Number	Ground cover: Percent Cover
P_GCOVER.P_STEMCNT	Text	Ground cover: Stem Count
P_INFO.P_TYPE	Text	Shrubs: Plant Type
P_INFO.FK_F_ORGAN	Text	Trees: Botanical Name
P_INFO.P_TYPE	Text	Trees: Plant Type
P_INFO.P_DBH	Number	Trees: Stem Diameter
P_INFO.P_HEIGHT	Number	Trees: Height
P_INFO.P_TYPE	Number	Plant Type
GRPTOFOR.GCONDITION	Likert scale	User group estimation: condition of this forest
GRPTOFOR.GCONSERVE	Likert scale	User group estimation: type of conservation measures

5 Results and Discussion

5.1 Comprehensive success factor synthesis

The following factor synthesis, comprising 260 concepts and indicators, forms the basis for the minimal synthesis used as the basis for the models:

1 Resource system
- 1.1 physical properties
  - a size of resource system
    - area
    - length
    - storage capacity
  - b system boundaries
    - congruence of biophysical and social boundaries
    - clarity
  - c location
    - accessibility
  - d storage characteristics
- 1.2 ecological properties
  - a sector
    - water-management
    - irrigation
    - forestry
    - energy
    - land use
    - pastures
    - agriculture
    - wildlife
  - b condition of resource
    - past
    - actual
    - future trend
  - c productivity of system
  - d equilibrium properties
  - e predictability of system dynamics
  - f biodiversity
- 1.3 human-related properties
  - a technology used
  - b value
  - c investments
    - human-constructed facilities

      d designated areas

         existence of designated areas for specific uses

         spatially explicit management

         protected areas, natural reserves (e.g., no-take or MPAs, marine protected areas)

         existence

         levels of protection

      e level of dependence on resource system

         primary livelihood

         occupational diversity

2 Resource units

  2.1 physical properties

      a number of units

      b distinctiveness

      c storage characteristics

  2.2 ecological properties

      a spatial and temporal distribution

      b mobility

      c predictability

      d regeneration

         growth or replacement rate

         global catch quotas

         seeding or restocking

         minimum sizes

         level of threat

      e interaction among RU

  2.3 human-related properties

      a economic value

      b appropriation/extraction

         user demand/amount extracted

         change in amount/levels of demand

         differences in individual appropriation levels

         trend levels of extraction

      c markings

      d technology used

      e manageability

         ease of harvesting

         ease of handling

3 Actors

  3.1 actors

      a properties of actors

        number of actors
        number of actors
        trend (in- or decrease)
        migration
        educational level
        years of education
        technical level of households
        income
        ethnicity
        common interests

b location
      distance to resource
      distance to market
      distance to local administration

3.2 group

a group composition
      heterogeneity
      number of castes
      GINI-coefficient
      number of ethnic parties
      percentage of migrants
      number of languages
      number of religions
      dependence on group members
      alternatives

b experience
      experience with self-organization, collective action, participation,
      SES in general
      history of use
      existence of organizations

c group boundaries
      clearly defined group boundaries
      individual requirements (e.g., auction, fees, land ownership)

d leadership
      existence of leader or leading group
      characteristics
         young
         connected to local traditional elite
         trusted by community
         experienced
         entrepreneurship

   e social capital
     shared norms
     past successful experiences/common history
     trust in community
     social cohesion
     community events
     conflicts between actors
     tradition
     future long-term commitment
   f economic properties
     economic condition of community
     income
       low levels of poverty
       wealth
     heterogeneity of endowments, homogeneity of identities and
       interests
     influence in local market
     percentage of households with electricity
   g information
     connections between information systems
     establish knowledge about collective action
     promotion of different approaches of information sharing
     expectations and limitations are openly stated
     existence of a common information base that is accessible
       and useful
       technological
       scientific
       social
       economic
       traditional knowledge
     economic evaluation of environmental as sets is a valuable
       information base
     community members are included in collection of scientific
       information
     knowledge of SES
       resource system
       resource units
       governance system
       actors
       external parties
   h communication

        properties
            effective
            open and transparent
        communication systems
        well-developed infrastructure
4 Governance system
    4.1   rule system
        a characteristics of rules
            existence of rules
            fair allocation of benefits
            simple, easy to understand
            flexible
            clear, transparent
            feedback systems
            local fit (adapted)
            capabilities to adapt to change
                flexibility and adaptability in future tasks
            access restrictions
                individual or community individual quotas
                match restrictions on harvests to regeneration resources
        b Monitoring
            environmental monitoring
                involvement of local actors
                guards
            social monitoring
                involvement of local actors
        c Enforcement
            enforcement of rules
                ease
                working
                self-enforcement
                level of rule-following
            sanctions
            existence
                adjudication
                availability
                low cost
                level of corruption
                accountability of monitors and other officials to actors
                graduated sanctions
            conflict-management

4.2 Rights
   a rights
      constitutional-choice rules
      collective-choice rules
      operational-choice rules
      property rights
         clear
         secure
         tenure
         participation
         appropriation
         access
         management
         exclusion
         provision
         legal certainty
         stability
         recognition of local rights by the state
   b participation
      level of self-organization
      level of participation
      effectiveness of participation
      locally devised model of participation
      legitimacy of decisions in participatory decisions
      percentage of group in decision-making processes or taking part
         in the process
      percentage of group in decision-making processes
      transparency of decision-making processes
         reading out loud protocols of last meeting
         accessibility of protocols, decisions, etc.
      accountability of decision makers to actors
      existence of institutions
      development of knowledge and capabilities
4.3 Organization
   a administration
      fit
         appropriate scale
         fit with traditional institutions and structures of community
      legitimacy of organization
      effective feedback systems
      age of organization

long-term management policy
social and technological capacities for monitoring, evaluating, enforcement of rules
organization provides information, about consulting, goals, decisions, and participatory possibilities
nested levels if larger systems
support through locally elected officials
clear goals

b Economic properties
working individual incentives
existence of financial resources and budget planning
cost of self-organization and institutional change
minimize effects on local economy
linkage between conservation and local economy
low-cost exclusion technology
local organization has financial resources and a labor force
working individual incentives
willingness of actors to invest money, labor, or time
expectations of actors concerning fits from self-organization
expectation that benefits accrue to actors

c relations to third parties
development of networks
government
NGOs
other group of actors

5.2 CPR

5.2.1 Descriptive statistics (key figures)

Table 7.5 Descriptive statistics for CPR

Variable name	Minimum	Maximum	Mean	Median	Variance	Std. dev.	Skewness	Kurtosis
F1 Resource size	−1	1	0.14	0.14	0.45	0.67	−0.49	−1.01
F2 Resource boundaries	−0.37	0.48	0.03	−0.05	0.05	0.23	0.72	−0.69
F3 Accessibility	−1	1	0.62	0.62	0.14	0.38	−1.21	2.5
F4 Initial ecological success	−1	1	0.08	0.12	0.19	0.44	−0.47	0.66
F5 Manageability	−0.87	0.57	−0.32	−0.44	0.12	0.35	0.61	−0.28
F6 Regeneration of RU	−1	1	−0.12	−0.05	0.25	0.5	0.19	−0.54
F7 Number of actors	−1	1	0.6	0.8	0.2	0.44	−1.9	3.49
F8 Group composition	−0.08	1	0.62	0.67	0.05	0.23	−0.91	0.38
F9 Social capital	−1	1	0.33	0.54	0.4	0.63	−0.65	−0.91
F10 Dependency on resource	−1	0.89	0.19	0.18	0.11	0.33	−0.68	1.57
F11 Dependency on group	−1	1	0.36	0.49	0.32	0.57	−1.13	0.42
F12 Group boundaries	−0.89	0.88	0.26	0.27	0.09	0.31	−1.54	3.46

F13 Participation of users	−0.87	0.84	0.17	0.23	0.16	0.39	−0.58	−0.48
F14 Legal certainty and legitimacy	−0.23	0.55	0.15	0.16	0.03	0.18	−0.02	−0.91
F15 Administration	−1	1	0.65	0.98	0.29	0.54	−1.87	2.8̇
F16 Information	−0.84	1	0.29	0.44	0.25	0.5	−0.62	−0.74
F17 Characteristics of rules	−1	0.68	0.18	0.23	0.12	0.34	−1.26	1.49
F18 Fairness	−1	1	0.41	0.65	0.31	0.56	−1.01	−0.06
F19 Control	−0.34	1	0.21	0.13	0.1	0.32	0.73	−0.23
F20 Compliance	−0.66	0.95	0.2	0.22	0.19	0.44	−0.2	−1.13
F21 Conflict management	−1	1	0.47	0.5	0.29	0.54	−1.49	1.67
F22 Exclusion	−0.69	1	0.38	0.42	0.12	0.35	−0.48	0.66
F23 Relations	−0.25	0.96	0.46	0.48	0.05	0.22	−0.63	0.65
F24 Capabilities to adapt to change	−0.9	1	−0.02	0.02	0.22	0.47	0.02	−1
F25 Success	−0.67	0.26	−0.17	−0.14	0.05	0.23	−0.2	−0.68

5.2.2   Histograms of success factors

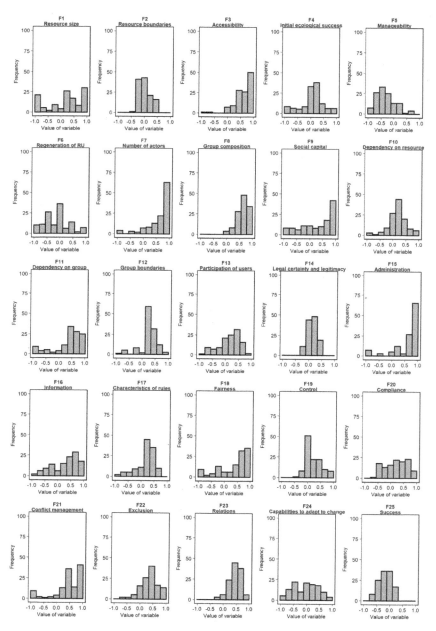

Figure 7.1   Histograms for the twenty-four success factors and ecological success for CPR

5.2.3 Correlation coefficients between success factors

Table 7.6 Correlation coefficients between success factors for CPR

	F1	F2	F3	F4	F5	F6	F7	F8	F9	F10	F11	F12	F13	F14	F15	F16	F17	F18	F19	F20	F21	F22	F23	F24
F2	−0.26**																							
F3		−0.24**																						
F4			0.18*																					
F5	0.31***	−0.30***																						
F6		−0.25**		0.18*																				
F7																								
F8	0.19*		−0.30***		0.22*																			
F9		0.23**	−0.29**			−0.25**																		
F10		0.21*	−0.23*			−0.29**			0.45***															
F11			−0.32***			−0.24**		0.31***	0.78***	0.40***														
F12		0.26**	−0.22*				0.18*																	
F13	0.39***	−0.28**						0.20*			0.42***	0.28**												
F14				0.28**	0.25**				0.36***		0.43***		0.36***											
F15		0.20*							0.19*															
F16		0.26**		−0.25**	0.28**	−0.21*				0.19*		0.18*	0.21*	0.27**										
F17			−0.30***							0.25**		0.27**		0.21*		0.20*								

Continued

Table 7.6 *Continued*

	F1	F2	F3	F4	F5	F6	F7	F8	F9	F10	F11	F12	F13	F14	F15	F16	F17	F18	F19	F20	F21	F22	F23	F24
F18	-0.24**	0.30***	-0.26**					0.24**	0.55***	0.36***	0.43***	0.23*		0.28**			0.20*							
F19	0.20*			-0.18*	0.37***								0.19*			0.47***		-0.19*						
F20	0.23*	-0.39***			0.33***	0.20*		0.21*			0.35***		0.65***	0.47***			0.21*							
F21				-0.22*	0.19*	0.33***		0.28**					0.19*			0.18*								
F22	0.21*				0.34***			0.30***	0.42***	0.37***	0.49***	0.34***	0.52***	0.36***		0.37***		0.30***	0.35***	0.24**				
F23	0.23**				0.27**	-0.19*		0.30***	0.28**		0.33***	0.21*	0.26*	0.34***		0.22*		0.28*				0.62***		
F24					0.33***			0.27**			0.32***		0.45***	0.36***	-0.20*	0.18*		0.27***		0.44***		0.46***	0.29***	
F25		0.20*		0.26**					0.32***	0.34***	0.39***		0.20*	0.48***			0.25***	0.42***		0.31***		0.20*	0.18*	0.31***

	F1	F2	F3	F4	F5	F6	F7	F8	F9	F10	F11	F12	F13	F14	F15	F16	F17	F18	F19	F20	F21	F22	F23	F24
F2																								
F3	-0.36***																							
F4			-0.16*																					
F5	-0.16*	-0.13*	0.14*																					
F6																								
F7	0.78***		-0.22***		-0.14*	0.24***																		
F8	0.19**		-0.14*																					
F9	0.21***																							
F10		0.17*	0.27***	-0.27***			0.13*	0.27***																
F11		0.17**				0.15*	-0.15*		0.45***															

	F1	F2	F3	F4	F5	F6	F7	F8	F9	F10	F11	F12	F13	F14	F15	F16	F17	F18	F19	F20	F21	F22	F23	F24
F12	0.18**						-0.20**		-0.14*	0.23***		0.29***	0.13*											
F13	0.20**	0.31***					-0.13*	-0.23***		0.30***		0.36***	0.43***	0.14*	0.28***									
F14	0.14*											0.26***	0.33***	0.16*	0.18**	0.36***								
F15	-0.13*						0.21***			-0.17**		-0.19**	-0.15*	0.18**										
F16	-0.13*	0.34***					-0.14*					0.18**	0.35***	0.16*	0.35***	0.30***								
F17	0.15*	0.13*					-0.17**	-0.15*		0.25***		0.44***	0.41***	0.21***	0.28***	0.46***	0.53***							
F18	0.18**	0.19**					-0.20**			0.30***		0.43***	0.38***		0.35***	0.38***	0.23***	-0.20**	0.22***	0.40***				
F19	-0.19**	0.22***								-0.13*				0.15*	-0.12*	0.26***	0.16**	0.18**	0.14*					
F20	0.34***						-0.17**					0.38***	0.50***	0.20**	0.13*	0.52***	0.35***	-0.20**	0.47***	0.44***	0.43***	0.18**		
F21																			0.14*	0.13*				
F22	0.24***						0.14*			0.14*		0.14*	0.19**		0.26***	0.14*	0.21***	0.20**	0.20**	0.23***				
F23	-0.26***						0.30***	0.20**		-0.23***					0.30***	0.28***			0.22***	-0.14*				
F24	0.18**						-0.27***			0.29***		0.13*	0.39***		0.24***	0.21***	-0.30***	0.18**	0.16*	0.38***	0.27***	0.14*	-0.27***	
F25	0.13*	0.25***					-0.12*	0.16*		-0.27***	0.24***	0.32***	0.49***		0.24***	0.46***	0.35***	0.16**	0.41***	0.38***	0.31***	0.23***	0.19**	

5.3.1 Descriptive statistics (key figures)

Table 7.7 Descriptive statistics for NIIS

Variable name	Minimum	Maximum	Mean	Median	Variance	Std. dev.	Skewness	Kurtosis
F1 Resource size	−1	1	0.53	0.74	0.3	0.55	−1.66	1.75
F2 Resource boundaries	−0.38	1	0.61	0.54	0.12	0.34	−0.48	−0.45
F3 Accessibility	−0.75	1	0.35	0.25	0.18	0.43	−0.07	−0.76
F4 Initial ecological success	−1	1	−0.34	−0.33	0.34	0.59	0.38	−0.7
F5 Manageability	−0.57	1	0.17	0.12	0.13	0.36	0.19	−0.73
F6 Regeneration of RU	−0.83	1	0	0	0.11	0.34	0.34	0.5
F7 Number of actors	−1	1	0.64	0.85	0.3	0.55	−2.21	3.74
F8 Group composition	−0.27	1	0.62	0.66	0.06	0.24	−1.21	1.51
F9 Social capital	−1	1	0.5	0.55	0.09	0.3	−1.54	4.14
F10 Dependency on resource	−0.83	0.75	0.18	0.2	0.1	0.31	−0.6	0.03
F11 Dependency on group	−1	0.9	0.15	0.16	0.19	0.43	−0.37	−0.78
F12 Group boundaries	−1	0.6	0.26	0.3	0.04	0.2	−2.2	8.48

F13 Participation of users	-0.75	0.83	0.33	0.39	0.08	0.27	-1.05	0.79
F14 Legal certainty and legitimacy	-0.6	0.89	0.26	0.28	0.06	0.25	-0.82	1.51
F15 Administration	-1	1	0.46	0.44	0.12	0.35	-0.35	0.18
F16 Information	-1	1	0.33	0.46	0.2	0.45	-0.88	-0.12
F17 Characteristics of rules	-0.75	0.79	0.37	0.44	0.07	0.26	-2.1	5.12
F18 Fairness	-1	1	0.35	0.48	0.15	0.39	-1.07	0.7
F19 Control	-0.8	0.87	0.11	0.13	0.1	0.32	-0.25	0.03
F20 Compliance	-0.91	0.83	0.18	0.26	0.15	0.38	-0.56	-0.48
F21 Conflict management	-1	1	0.38	0.38	0.24	0.49	-0.75	-0.25
F22 Exclusion	-1	1	-0.24	-0.33	0.29	0.54	0.2	-0.59
F23 Relations	-1	1	0.05	0.1	0.14	0.37	-0.55	0.46
F24 Capabilities to adapt to change	-1	1	-0.19	-0.25	0.19	0.44	0.54	-0.28
F25 Success	-0.71	0.99	0.28	0.33	0.15	0.39	-0.41	-0.64

5.3.2 Histograms of success factors

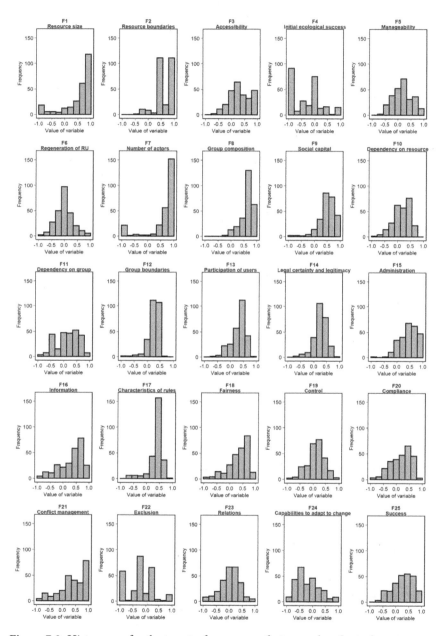

Figure 7.2 Histograms for the twenty-four success factors and ecological success for NIIS

Table 7.8 Correlation coefficients between success factors for NIIS

	F1	F2	F3	F4	F5	F6	F7	F8	F9	F10	F11	F12	F13	F14	F15	F16	F17	F18	F19	F20	F21	F22	F23	F24
F2																								
F3	-0.36***																							
F4			-0.16*																					
F5	-0.16*		0.14*	-0.13*																				
F6																								
F7	0.78***		-0.22***		-0.14*																			
F8	0.19**		-0.14*																					
F9	0.21***						0.24***																	
F10		0.17**	0.27***	-0.27***			0.13*		0.27**															
F11		0.17**	0.27***			0.15*	-0.15*		0.45***															
F12	0.18**			-0.20**		-0.14*	0.23***		0.29***	0.13*														
F13	0.20**	0.31***		-0.13*	-0.23***		0.30***		0.36***	0.43***	0.14*	0.28***												
F14		0.14*							0.26***	0.33***	0.16*	0.18**	0.36***											
F15	-0.13*			0.21***			-0.17*	-0.17**	0.21***	-0.19**	-0.15*	0.18**												
F16	-0.13*	0.34***		-0.14*					0.18**	0.35***	0.16**	0.16*	0.35***	0.30***										
F17	0.15*	0.13*		-0.17*	-0.15*		0.25***		0.44***	0.41***	0.21***	0.28***	0.46***	0.53***		0.40***								

Continued

Table 7.8 Continued

	F1	F2	F3	F4	F5	F6	F7	F8	F9	F10	F11	F12	F13	F14	F15	F16	F17	F18	F19	F20	F21	F22	F23	F24
F18	0.18**	0.19**		−0.20**			0.30***		0.43***	0.38***		0.35***	0.38***	0.23***	−0.20***	0.22***	0.40***							
F19	−0.19**	0.22***					−0.13*				0.15*	−0.12*	0.26***	0.16**	0.18**	0.14*								
F20		0.34***		−0.17**					0.38***	0.50***	0.20**	0.13*	0.52***	0.35***	−0.20**	0.47***	0.44***	0.43***	0.18**					
F21																0.14*		0.13*						
F22		0.24***					0.14*		0.14*	0.19**			0.26***		0.14*	0.21***	0.20**	0.20**		0.23***				
F23	−0.26***			0.30***	0.20**		−0.23***							0.30***	0.28***				0.22***		−0.14*			
F24	0.18**			−0.27***			0.29***		0.13*	0.39***		0.24***	0.21***		−0.30***	0.18*	0.16*	0.38***		0.27***	0.14*		−0.27***	
F25	0.13*		0.25***	−0.12*	0.16*	−0.27***	0.24***		0.32***	0.49***		0.24***	0.46***	0.35***	0.16*	0.41***	0.38***	0.38***		0.31***	0.14*	0.23***		0.19**

5.4 IFRI

5.4.1 Descriptive statistics (key figures)

Table 7.9 Descriptive statistics for IFRI

Variable name	Minimum	Maximum	Mean	Median	Variance	Std. dev.	Skewness	Kurtosis
F1 Resource size	−1	1	0.19	0.19	0.2	0.45	−0.17	−0.43
F2 Resource boundaries	−1	1	−0.59	−1	0.36	0.6	1.35	0.74
F3 Accessibility	−1	1	0.15	0.15	0.19	0.44	−0.08	−0.65
F4 Initial ecological success	−0.74	0.55	0.08	0.08	0.02	0.14	−0.59	5.59
F5 Manageability	−1	1	0.29	0.36	0.15	0.39	−0.92	1.17
F6 Regeneration of RU	−0.91	1	0.27	0.23	0.08	0.29	0.77	1.53
F7 Number of actors	−0.97	0.9	−0.1	−0.1	0.11	0.33	0.08	0.15
F8 Group composition	−1	0.68	−0.09	−0.06	0.09	0.3	−0.4	−0.02
F9 Social capital	−1	1	−0.07	−0.07	0.11	0.33	−0.1	−0.41
F10 Dependency on resource	−0.88	0.32	−0.15	−0.13	0.03	0.17	−0.96	2.73
F11 Dependency on group	−1	1	−0.12	−0.12	0.07	0.27	0.28	1.39
F12 Group boundaries	−1	1	−0.03	−0.04	0.18	0.42	0.2	0.17

Continued

Table 7.9 *Continued*

Variable name	Minimum	Maximum	Mean	Median	Variance	Std. dev.	Skewness	Kurtosis
F13 Participation of users	−1	1	−0.02	0.04	0.14	0.38	−0.66	0.06
F14 Legal certainty and legitimacy	−1	1	0.2	0.24	0.09	0.3	−0.57	1.72
F15 Administration	−1	1	−0.03	−0.03	0.11	0.33	−0.01	2.11
F16 Information	−1	0.9	0.04	0.12	0.15	0.39	−0.75	0.34
F17 Characteristics of rules	−0.84	0.76	−0.13	−0.15	0.06	0.25	0.84	1.55
F18 Fairness	−1	1	0.2	0.19	0.13	0.36	−0.1	0.33
F19 Control	−1	1	−0.12	−0.12	0.1	0.32	0.42	1.95
F20 Compliance	−0.56	1	0.24	0.26	0.08	0.29	0.06	−0.02
F21 Conflict management	−1	1	0.11	0.11	0.16	0.4	0.42	0.46
F22 Exclusion	−1	0.8	0.03	0.04	0.1	0.32	−0.1	−0.32
F23 Relations	−0.75	0.75	−0.03	−0.02	0.06	0.24	−0.02	0.8
F24 Capabilities to adapt to change	−0.68	0.61	−0.25	−0.28	0.05	0.22	1.18	2.1
F25 Success	−0.74	0.79	0.11	0.14	0.06	0.25	−0.31	0.18

5.4.2 Histograms of success factors

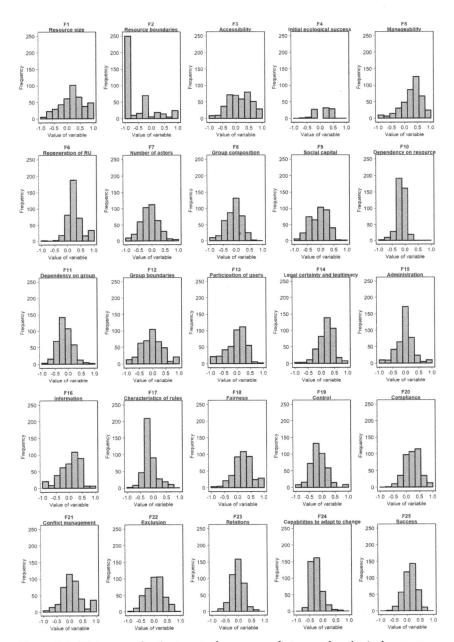

Figure 7.3 Histograms for the twenty-four success factors and ecological success for IFRI

5.4.3 Correlation coefficients between success factors for IFRI

Table 7.10 Correlation coefficients between success factors

	F1	F2	F3	F4	F5	F6	F7	F8	F9	F10	F11	F12	F13	F14	F15	F16	F17	F18	F19	F20	F21	F22	F23	F24
F2	−0.31***																							
F3	−0.24***																							
F4	−0.15**																							
F5	−0.30***	0.13**	0.49***	0.12*																				
F6	0.17***	−0.12*	−0.14**		−0.14**																			
F7	0.23***				−0.16**	0.11*																		
F8	−0.10*	0.10*	0.14**		0.22***	−0.21***	−0.16***																	
F9		0.11*			0.16**		−0.16**																	
F10	−0.27***	0.27***	0.18***		0.12*	−0.42***	0.11*		0.18***															
F11	−0.26***		0.15**		0.12*	−0.15**																		
F12					0.12*				0.26***															
F13		0.21***			0.11*				0.41***	0.29***		0.16***												
F14		0.15**							0.18***	0.19***		0.20***	0.16***											
F15	0.23***					−0.15**			0.25***			0.11*	0.26***	0.17***										
F16	−0.18***	0.28***		0.12*	0.22***	−0.17***			0.48***	0.31***		0.16**	0.47***		0.25***									
F17				−0.22***		0.11*						0.15*				−0.15**								

F18	0.18***			0.23***			0.30***	−0.13*			0.18***	0.20***			0.26***	0.24***		
F19				0.11*			0.11*			0.11*	0.13**			0.27***	0.29***	0.21***		0.18***
F20	0.10*	−0.13**	−0.12*	0.16**			0.25***			0.15**	0.22***	0.14**	0.13**	0.33***	0.23***	0.25***	0.19***
F21			−0.11*				−0.20***			−0.11*					−0.28***	0.30***	−0.18***
F22	0.12*				0.33***	−0.15**	−0.13**	0.30***				0.34***			−0.18***	0.11*
F23	−0.12*	0.24***	0.11*		−0.28***		0.38***	0.16**		0.34***	0.29***	0.20***	0.18***	0.52***		0.18***	0.21***	0.30*** −0.11* −0.16**
F24		0.14**		0.35***	−0.11*	0.12*	−0.12*	0.10*			0.15**						0.18***
F25	−0.20***	0.21***	0.31***	0.13*		0.12*	0.19***		0.17***	0.24***		0.21***				0.13*

5.5 Combined model

5.5.1 Descriptive statistics (key figures)

Table 7.11 Descriptive statistics for the combined model

Variable name	Minimum	Maximum	Mean	Median	Variance	Std. dev.	Skewness	Kurtosis
F1 Resource size	−1	1	0.29	0.36	0.3	0.55	−0.69	−0.32
F2 Resource boundaries	−1	1	−0.1	−0.11	0.53	0.73	0.04	−1.37
F3 Accessibility	−1	1	0.29	0.27	0.21	0.45	−0.2	−0.62
F4 Initial ecological success	−1	1	−0.06	0.08	0.19	0.44	−0.79	0.83
F5 Manageability	−1	1	0.15	0.19	0.18	0.43	−0.38	−0.39
F6 Regeneration of RU	−1	1	0.12	0.13	0.14	0.38	−0.09	0.77
F7 Number of actors	−1	1	0.25	0.23	0.32	0.56	−0.3	−0.96
F8 Group composition	−1	1	0.25	0.25	0.2	0.45	−0.31	−0.83
F9 Social capital	−1	1	0.18	0.19	0.22	0.47	−0.22	−0.7
F10 Dependency on resource	−1	0.89	0.01	−0.04	0.09	0.3	0.21	0.22
F11 Dependency on group	−1	1	0.04	0.02	0.18	0.43	0.15	−0.34
F12 Group boundaries	−1	1	0.11	0.2	0.14	0.37	−0.59	0.56
F13 Participation of users	−1	1	0.12	0.19	0.15	0.38	−0.76	0.19

F14 Legal certainty and legitimacy	−1	1	0.21	0.24	0.07	0.27	−0.6	1.8
F15 Administration	−1	1	0.24	0.16	0.22	0.47	−0.07	−0.26
F16 Information	−1	1	0.18	0.25	0.2	0.45	−0.54	−0.25
F17 Characteristics of rules	−1	0.79	0.09	0.08	0.12	0.35	−0.15	−0.91
F18 Fairness	−1	1	0.28	0.32	0.17	0.41	−0.54	0.11
F19 Control	−1	1	0	−0.01	0.12	0.35	0.2	0.53
F20 Compliance	−0.91	1	0.22	0.25	0.12	0.35	−0.35	−0.2
F21 Conflict management	−1	1	0.26	0.19	0.23	0.48	−0.25	−0.41
F22 Exclusion	−1	1	−0.01	0.03	0.21	0.46	−0.31	−0.09
F23 Relations	−1	1	0.07	0.06	0.11	0.33	−0.15	0.32
F24 Capabilities to adapt to change	−1	1	−0.2	−0.25	0.13	0.36	0.81	0.69
F25 Success	−0.74	0.99	0.13	0.13	0.11	0.33	0.05	−0.29

5.5.2 Histograms of success factors

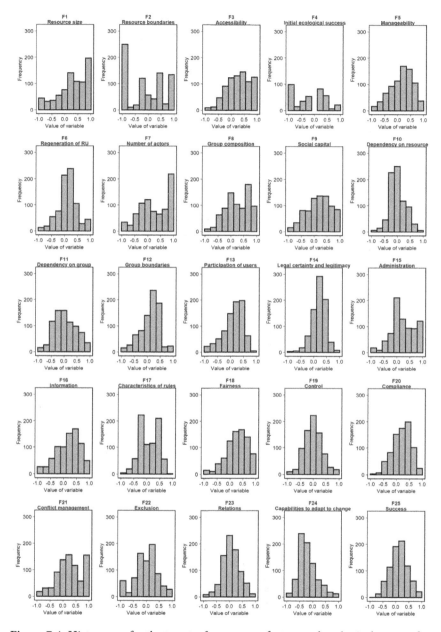

Figure 7.4 Histograms for the twenty-four success factors and ecological success for the combined dataset

5.5.3 Correlation coefficients between success factors

Table 7.12 Correlation coefficients between success factors for the combined model

	F1	F2	F3	F4	F5	F6	F7	F8	F9	F10	F11	F12	F13	F14	F15	F16	F17	F18	F19	F20	F21	F22	F23	F24
F2																								
F3	-0.18***	0.21***																						
F4	-0.13***	-0.32***	-0.08*																					
F5	-0.09**	-0.11**	0.09*																					
F6		-0.30***	-0.21***	0.12***	0.12***																			
F7	0.46***	0.45***	0.10**	-0.22***	-0.27***	-0.24***																		
F8	0.18***	0.59***	0.25***	-0.23***	-0.19***	-0.36***	0.49***																	
F9	0.11**	0.48***	0.11**	-0.22***		-0.28***	0.40***	0.45***																
F10		0.49***	0.28***	-0.31***	-0.12***	-0.40***	0.41***	0.46***	0.51***															
F11	-0.08*	0.28***	0.18***		-0.15***	-0.23***	0.20***	0.35***	0.53***	0.35***														
F12	0.11**	0.35***	0.09*	-0.15***		-0.14***	0.29***	0.29***	0.38***	0.29***	0.17***													
F13	0.27***	0.42***	0.10**	-0.19***		-0.15***	0.32***	0.31***	0.47***	0.43***	0.25***	0.32***												
F14		0.16***							0.22**	0.21***	0.10**	0.18***	0.25***											
F15	0.11**	0.42***	0.19***		-0.26***	-0.28***	0.31***	0.48***	0.39***	0.29***	0.31***	0.26***	0.31***	0.13***										
F16		0.39***	0.08*	-0.20***		-0.24***	0.17***	0.27***	0.39***	0.40***	0.21***	0.25***	0.45***	0.14***	0.25***									
F17	0.19***	0.49***	0.12***	-0.33***	-0.15***	-0.20***	0.49***	0.50***	0.45***	0.48***	0.26***	0.37***	0.37***	0.22***	0.34***	0.27***								

Continued

Table 7.12 *Continued*

	F1	F2	F3	F4	F5	F6	F7	F8	F9	F10	F11	F12	F13	F14	F15	F16	F17	F18	F19	F20	F21	F22	F23	F24
F18	0.09*	0.19***		-0.11**			0.23***	0.19***	0.44***	0.28***	0.21***	0.26***	0.26***	0.14***	0.18***	0.24***	0.26***							
F19		0.31***	0.14***	-0.09*	-0.10**	-0.14***	0.20***	0.28***	0.25***	0.25***	0.26***	0.19***	0.26***	0.21***	0.37***	0.32***	0.26***	0.14***						
F20	0.08*					0.14***										0.27***	0.18***	0.27***	0.13***					
F21	0.11**	0.19***		-0.16***	-0.14***		0.24***	0.26***	0.15***	0.19***	0.11**	0.17***	0.33***	0.24***	0.15***	0.07	0.32***	0.08*	0.10*					
F22		-0.13***	0.09**	0.16***	-0.16***	-0.08**	0.08*	-0.09**		0.15***	0.15***			0.13***				0.09**		0.13***				
F23	-0.14***	0.19***	0.16***	0.19***	-0.15***	-0.20***		0.29***	0.24***	0.23***	0.29***	0.27***	0.20***	0.17***	0.34***	0.31***	0.13***	0.15***	0.31***			0.14***		
F24	0.09*		0.17***	-0.11**			0.20***	0.10*	0.17***	0.26***	0.17***	0.16***	0.20***	0.07		0.14***	0.14***	0.30***	0.08*	0.24***	0.14***	0.13***	0.07	
F25	0.08*	0.23***		-0.15***	0.23***		0.16***		0.23***	0.29***		0.14***	0.31***			0.15***	0.24***	0.18***	0.19***	0.19***	0.14***	-0.07*	-0.15***	

References

Achard, Frédéric, René Beuchle, Philippe Mayaux, Hans-Jürgen Stibig, Catherine Bodart, Andreas Brink, Silvia Carboni et al. 2014. "Determination of Tropical Deforestation Rates and Related Carbon Losses from 1990 to 2010." *Global Change Biology* 20 (8): 2540–54. https://doi.org/10.1111/gcb.12605.

Acheson, James M. 1987. "The Lobster Fiefs Revisited: Economic and Ecological Effects of Territoriality in the Maine Lobster Industry." In McCay and Acheson 1987, 37–65.

Addison, Jane, and Romy Greiner. 2016. "Applying the Social-Ecological Systems Framework to the Evaluation and Design of Payment for Ecosystem Service Schemes in the Eurasian Steppe." *Biodiversity and Conservation* 25 (12): 2421–40. https://doi.org/10.1007/s10531-015-1016-3.

Aglionby, Julia. 2014. *The Governance of Commons in National Parks: Plurality and Purpose.* Newcastle: Newcastle University.

Agrawal, Arun. 2001. "Common Property Institutions and Sustainable Governance of Resources." *World Development* 29 (10): 1649–72.

Agrawal, Arun. 2002. "Common Resources and Institutional Sustainability." In Ostrom, Dietz, Dolšak, Stern, Stonich, and Weber 2002, 41–85.

Agrawal, Arun. 2007. "Forests, Governance, and Sustainability: Common Property Theory and Its Contributions." *International Journal of the Commons* 1 (1): 111–36.

Agrawal, Arun, and Ashwini Chhatre. 2006. "Explaining Success on the Commons: Community Forest Governance in the Indian Himalaya." *World Development* 34 (1): 149–66. https://doi.org/10.1016/j.worlddev.2005.07.013.

Agrawal, Arun, and Clark C. Gibson. 1999. "Enchantment and Disenchantment: The Role of Community in Natural Resource Conservation." *World Development* 27 (4): 629–49. https://doi.org/10.1016/S0305-750X(98)00161-2.

Agrawal, Arun, and Gautam N. Yadama. 1997. "How Do Local Institutions Mediate Market and Population Pressures on Resources? Forest Panchayats in Kumaon, India." *Development and Change* 28 (3): 435–65.

Alcamo, Joseph, Martina Flörke, and Michael Märker. 2007. "Future Long-Term Changes in Global Water Resources Driven by Socio-Economic and Climatic Changes." *Hydrological Sciences Journal* 52 (2): 247–75. https://doi.org/10.1623/hysj.52.2.247.

Alfano, Geraldine, and Gerald Marwell. 1980. "Experiments on the Provision of Public Goods by Groups III: Nondivisibility and Free Riding in 'Real' Groups." *Social Psychology Quarterly* 43 (3): 300–309.

Alpaydin, Ethem. 2010. *Introduction to Machine Learning.* 2nd ed. Adaptive computation and machine learning. Cambridge, Mass. MIT Press.

Altieri, Miguel A. 2002. "Agroecology: The Science of Natural Resource Management for Poor Farmers in Marginal Environments." *Agriculture, Ecosystems and Environment* 93 (1-3): 1–24. https://doi.org/10.1016/S0167-8809(02)00085-3.

Anderies, John M., and Marco A. Janssen. 2013. "Sustaining the Commons." Accessed March 26, 2014. https://csid.asu.edu/publications/sustaining-commons.

Anderies, John M., Marco A. Janssen, and Elinor Ostrom. 2004. "A Framework to Analyze the Robustness of Social-Ecological Systems from an Institutional Perspective." *Ecology and Society* 9 (1): 18.

Anderies, John M., Marco A. Janssen, and Edella Schlager. 2016. "Institutions and the Performance of Coupled Infrastructure Systems." *International Journal of the Commons* 10 (2): 495. https://doi.org/10.18352/ijc.651.

Andersson, Krister P., and Arun Agrawal. 2011. "Inequalities, Institutions, and Forest Commons." *Global Environmental Change* 21 (3): 866–75. https://doi.org/10.1016/j.gloenvcha.2011.03.004.

Andreoni, James. 1988. "Why Free Ride? Strategies and Learning in Public Goods Experiments." *Journal of Public Economics* 37 (3): 291–304.

Armitage, Derek R., Ryan Plummer, Fikret Berkes, Robert I. Arthur, Anthony T. Charles, Iain J. Davidson-Hunt, Alan P. Diduck et al. 2009. "Adaptive Co-Management for Social-Ecological Complexity." *Frontiers in Ecology and the Environment* 7 (2): 95–102. https://doi.org/10.1890/070089.

Arnstein, Sherry R. 1969. "A Ladder of Citizen Participation." *Journal of the American Institute of Planners* 35 (4): 216–24. https://doi.org/10.1080/01944366908977225.

Atkinson, Quentin D., and Pierrick Bourrat. 2011. "Beliefs About God, the Afterlife and Morality Support the Role of Supernatural Policing in Human Cooperation." *Evolution and Human Behavior* 32: 41–49.

Axelrod, Robert. 1984/2000. *Die Evolution der Kooperation*. München: Oldenbourg.

Axelrod, Robert, and William D. Hamilton. 1981. "The Evolution of Cooperation." *Science* 211 (4489): 1390–96. https://doi.org/10.1126/science.7466396.

Backhaus, Klaus, Bernd Erichson, Wulff Plinke, and R Weiber. 2008. *Multivariate Analysemethoden: Eine anwendungsorientierte Einführung, 12. Auflage*. Heidelberg: Springer.

Backhaus, Klaus, Bernd Erichson, and Rolf Weiber. 2013. *Fortgeschrittene Multivariate Analysemethoden: Eine Anwendungsorientierte Einführung*. 2., überarb. und erw. Aufl. Lehrbuch. Berlin [u.a.]: Springer Gabler.

Baland, Jean-Marie, and Jean-Philippe Platteau. 1996. *Halting Degradation of Natural Resources: Is There a Role for Rural Communities?* Oxford: Clarendon Press.

Ban, Natalie C., Tammy E. Davies, Stacy E. Aguilera, Cassandra Brooks, Michael Cox, Graham Epstein, Louisa S. Evans, Sara M. Maxwell, and Mateja Nenadovic. 2017. "Social and Ecological Effectiveness of Large Marine Protected Areas." *Global Environmental Change* 43: 82–91. https://doi.org/10.1016/j.gloenvcha.2017.01.003.

Barclay, Pat. 2004. "Trustworthiness and Competitive Altruism Can Also Solve the 'Tragedy of the Commons'." *Evolution and Human Behavior* 25 (4): 209–20.

Barclay, Pat. 2008. "Enhanced Recognition of Defectors Depends on Their Rarity." *Cognition* 107: 817–28.

Barkow, Jerome H., Leda Cosmides, and John Tooby, eds. 1992. *The Adapted Mind: Evolutionary Psychology and the Generation of Culture*. Oxford: Oxford University Press.

Basurto, Xavier, Stefan Gelcich, and Elinor Ostrom. 2013. "The Social-Ecological System Framework as a Knowledge Classificatory System for Benthic Small-Scale Fisheries." *Global Environmental Change* 23 (6): 1366–80. https://doi.org/10.1016/j.gloenvcha.2013.08.001.

Bateson, Melissa, Daniel Nettle, and Gilbert N. Roberts. 2006. "Cues of Being Watched Enhance Cooperation in a Real-World Setting." *Biology Letters* 2: 412–14.

Baur, Ivo, and Claudia R. Binder. 2013. "Adapting to Socioeconomic Developments by Changing Rules in the Governance of Common Property Pastures in the Swiss Alps." *Ecology and Society* 18 (4): 60–75. https://doi.org/10.5751/ES-05689-180460.

Baur, Ivo, Karina Liechti, and Claudia R. Binder. 2014. "Why Do Individuals Behave Differently in Commons Dilemmas? The Case of Alpine Farmers Using Common Property Pastures in Grindelwald, Switzerland." *International Journal of the Commons* 8 (2): 657. https://doi.org/10.18352/ijc.469.

Berkes, Fikret. 1986. "Local-Level Management and the Commons Problem: A Comparative Study of Turkish Coastal Fisheries." *Marine Policy* 10: 215–29.

Berkes, Fikret. 1987. "Common-Property Resource Management and Cree Indian Fisheries in Subarctic Canada." In McCay and Acheson 1987, 66–91.

Berkes, Fikret. 1992. "Success and Failure in Marine Coastal Fisheries of Turkey." In Bromley, Feeny, Peters, Gilles, Oakerson, Runge, and Thomson, eds. 1992, 161–82.

Berkes, Fikret. 2007. "Community-Based Conservation in a Globalized World." *Proceedings of the National Academy of Sciences of the United States of America* 104 (39): 15188–93.

Berkes, Fikret. 2009. "Evolution of Co-Management: Role of Knowledge Generation, Bridging Organizations and Social Learning." *Journal of Environmental Management* 90 (5): 1692–1702. https://doi.org/10.1016/j.jenvman.2008.12.001.

Berkes, Fikret, Johan Colding, and Carl Folke. 2003. *Navigating Social-Ecological Systems: Building Resilience for Complexity and Change.* Cambridge, New York: Cambridge University Press.

Berkes, Fikret, Robin Mahon, Patrick McConney, Richard Pollnac, and Robert S. Pomeroy. 2001. *Managing Small-Scale Fisheries: Alternative Directions and Methods.* Ottawa: International Development Research Centre.

Biermann, Frank, Norichika Kanie, and Rakhyun E. Kim. 2017. "Global Governance by Goal-Setting: The Novel Approach of the UN Sustainable Development Goals." *Current Opinion in Environmental Sustainability* 26-27: 26–31. https://doi.org/10.1016/j.cosust.2017.01.010.

Binder, Claudia R., Pieter Bots, Jochen Hinkel, and Claudia Pahl-Wostl. 2013. "Comparison of Frameworks for Analysing Social-Ecological Systems." *Ecology and Society* 18 (4): 26–44. http://www.ecologyandsociety.org/vol18/iss4/art26/.

Birkhofer, Klaus, Eva Diehl, Jesper Andersson, Johan Ekroos, Andrea Früh-Müller, Franziska Machnikowski, Viktoria L. Mader et al. 2015. "Ecosystem Services - Current Challenges and Opportunities for Ecological Research." *Frontiers in Ecology and Evolution* 2: 413. https://doi.org/10.3389/fevo.2014.00087.

Bischoff, Ivo. 2007. "Institutional Choice Vs Communication in Social Dilemmas - an Experimental Approach." *Journal of Economic Behavior and Organization* 62: 20–36.

Black, Maggie, and Jannet King. 2009. *The Atlas of Water: Mapping the World's Most Critical Resource.* London: Earthscan.

Bloom, Paul, and Tim P. German. 2000. "Two Reasons to Abandon the False Belief Task as a Test of Theory of Mind." *Cognition* 77 (1): B25-B31.

Blythe, J L. 2015. "Resilience and Social Thresholds in Small-Scale Fishing Communities." *Sustainability Science* 10 (1): 157–65. https://doi.org/10.1007/s11625-014-0253-9.

Blythe, Jessica, Philippa Cohen, Hampus Eriksson, Joshua Cinner, Delvene Boso, Anne-Maree Schwarz, and Neil Andrew. 2017. "Strengthening Post-Hoc Analysis of Community-Based Fisheries Management through the Social-Ecological Systems Framework." *Marine Policy* 82: 50–58. https://doi.org/10.1016/j.marpol.2017.05.008.

Bochet, Olivier, Talbot Page, and Louis Putterman. 2006. "Communication and Punishment in Contribution Experiments." *Journal of Economic Behavior and Organization* 60 (1): 11–26. https://doi.org/10.1016/j.jebo.2003.06.006.

Bochet, Olivier, and Louis Putterman. 2009. "Not Just Babble: Opening the Black Box of Communication in a Voluntary Contribution Experiment." *European Economic Review* 53: 309–26.

Boesch, Christophe. 2001. "Cooperative Hunting Roles among Tai Chimpanzees." *Human Nature* 13 (1): 27–46.

Böhringer, Christophe, and Patrick E. Jochem. 2007. "Measuring the Immeasurable - a Survey of Sustainability Indices." *Ecological Economics* 63: 1–8.

Bortz, Jürgen, and Christof Schuster. 2010. *Statistik Für Human- Und Sozialwissenschaftler.* 7., vollst. überarb. u. erw. Aufl. Springer-Lehrbuch. Berlin: Springer.

Boyd, Heather, and Anthony Charles. 2006. "Creating Community-Based Indicators to Monitor Sustainability of Local Fisheries." *Ocean & Coastal Management* 49 (5-6): 237–58. https://doi.org/10.1016/j.ocecoaman.2006.03.006.

Boyd, Robert, Herbert Gintis, Samuel Bowles, and Peter J. Richerson. 2003. "The Evolution of Altruistic Punishment." *Proceedings of the National Academy of Sciences of the United States of America* 100 (6): 3531–35.

Breiman, Leo. 2001. "Random Forests." *Machine Learning* 45 (1): 5–32. https://doi.org/10.1023/A:1010933404324.

Brockhurst, Michael A., Angus Buckling, Dan Racey, and Andy Gardner. 2008. "Resource Supply and the Evolution of Public-Goods Cooperation in Bacteria." *BMC Biology* 6: 20–26.

Brockhurst, Michael A., Michael E. Hochberg, Thomas Bell, and Angus Buckling. 2006. "Character Displacement Promotes Cooperation in Bacterial Biofilms." *Current Biology* 16 (20): 2030–34.

Brooks, J S., K A. Waylen, and Monique Borgerhoff Mulder. 2012. "How National Context, Project Design, and Local Community Characteristics Influence Success in Community-Based Conservation Projects." *Proceedings of the National Academy of Sciences of the United States of America* 109 (52): 21265–70. https://doi.org/10.1073/pnas.1207141110.

Brown-Kruse, Jamie, and David Hummels. 1993. "Gender Effects in Laboratory Public Goods Contribution: Do Individuals Put Their Money Where Their Mouth Is?" *Journal of Economic Behavior and Organization* 22 (3): 255–67.

Brundtland, Gru, Mansour Khalid, Susanna Agnelli, Sali Al-Athel, Bernard Chidzero, Lamina Fadika, Volker Hauff et al. 1987. *Our Common Future.* Oxford: Oxford University Press. http://www.bne-portal.de/fileadmin/unesco/de/Downloads/Hintergrundmaterial_international/Brundtlandbericht.File.pdf?linklisted=2812.

Bshary, Redouan, and Alexandra S. Grutter. 2002. "Asymmetric Cheating Opportunities and Partner Control in a Cleaner Fish Mutualism." *Animal Behaviour* 63 (3): 547–55. https://doi.org/10.1006/anbe.2001.1937.

Bshary, Redouan, and Alexandra S. Grutter. 2006. "Image Scoring and Cooperation in a Cleaner Fish Mutualism." *Nature* 441: 975–78.

Buchan, Nancy R., Rachel T. A. Croson, and Robyn M. Dawes. 2002. "Swift Neighbors and Persistent Strangers: A Cross-Cultural Investigation of Trust and Reciprocity in Social Exchange." *American Journal of Sociology* 108 (1): 168–206.

Buchan, Nancy R., Gianluca Grimalda, Rick Wilson, Marilynn Brewer, Enrique Fatas, and Margaret Foddy. 2009. "Globalization and Human Cooperation." *Proceedings of the National Academy of Sciences of the United States of America* 106 (11): 4138–42.

Bundesregierung. 2015. "Perspektiven für Deutschland: Unsere Strategie für eine nachhaltige Entwicklung." Accessed October 30, 2017. https://www.bundesregierung. de/Webs/Breg/DE/Themen/Nachhaltigkeitsstrategie/_node.html.

Burnham, Terence C., and Dominic D. P. Johnson. 2005. "The Biological and Evolutionary Logic of Human Cooperation." *Analyse und Kritik* 27 (1): 113–35.

Cadsby, Charles B., and Elizabeth Maynes. 1998a. "Choosing Between a Socially Efficient and Free-Riding Equilibrium: Nurses versus Economics and Business Students." *Journal of Economic Behavior and Organization* 37: 183–92.

Cadsby, Charles B., and Elizabeth Maynes. 1998b. "Gender and Free Riding in a Threshold Public Goods Game: Experimental Evidence." *Journal of Economic Behavior and Organization* 34: 603–20.

Cameron, Lisa. 1995. "Raising the Stakes in the Ultimatum Game: Experimental Evidence from Indonesia." *Economic Inquiry* 37 (1): 47–59.

Cardenas, Juan-Camilo, John Stranlund, and Cleve Willis. 2000. "Local Environmental Control and Institutional Crowding-Out." *World Development* 28 (10): 1719–33.

Cardenas, Samuel. 2000. "How Do Groups Solve Local Commons Dilemmas? Lessons from Experimental Economics in the Field." *Environment, Development and Sustainability* 2 (3-4): 305–22.

Carlsson, Lars, and Fikret Berkes. 2005. "Co-Management: Concepts and Methodological Implications." *Journal of Environmental Management* 75 (1): 65–76. https://doi.org/ 10.1016/j.jenvman.2004.11.008.

Carpenter, Jeffrey P. 2007. "Punishing Free-Riders: How Group Size Affects Mutual Monitoring and the Provision of Public Goods." *Games and Economic Behavior* 60 (1): 31–51. https://doi.org/10.1016/j.geb.2006.08.011.

Carpenter, Jeffrey P., Samuel Bowles, Herbert Gintis, and Sung-Ha Hwang. 2009. "Strong Reciprocity and Team Production: Theory and Evidence." *Journal of Economic Behavior and Organization* 71 (2): 221–32.

Cars, Otto, Anna Hedin, and Andreas Heddini. 2011. "The Global Need for Effective Antibiotics - Moving Towards Concerted Action." *Drug Resistance Updates* 14 (2): 68–69. https://doi.org/10.1016/j.drup.2011.02.006.

Casari, Marco. 2003. "Decentralized Management of Common Property Resources: Experiments with a Centuries-Old Institution." *Journal of Economic Behavior and Organization* 51 (2): 217–47. https://doi.org/10.1016/S0167-2681(02)00098-7.

Cavalcanti, Carina, Felix Schläpfer, and Bernhard Schmid. 2010. "Public Participation and Willingness to Cooperate in Common-Pool Resource Management: A Field Experiment with Fishing Communities in Brazil." *Ecological Economics* 69: 613–22.

Cesarini, David, Christopher T. Dawes, James H. Fowler, Magnus Johannesson, Paul Lichtenstein, and Björn Wallace. 2008. "Heritability of Cooperative Behavior in the Trust Game." *Proceedings of the National Academy of Sciences of the United States of America* 105 (10): 3721–26.

Chapman, Gretchen B., and Eric J. Johnson. 2002. "Incorporating the Irrelevant: Anchors in Judgments of Belief and Value." In *Heuristics and Biases: The Psychology of Intuitive Judgment*, edited by Thomas Gilovich, Dale Griffin, and Daniel Kahneman, 120–38. Cambridge: Cambridge University Press.

Chaudhuri, Ananish. 2011. "Sustaining Cooperation in Laboratory Public Goods Experiments: A Selective Survey of the Literature." *Experimental Economics* 14 (1): 47–83. 10.1007\s10683-010-9257-1.

Chhatre, Ashwini, and Arun Agrawal. 2008. "Forest Commons and Local Enforcement." *Proceedings of the National Academy of Sciences of the United States of America* 105 (36): 13286–91. https://doi.org/10.1073/pnas.0803399105.

Cifdaloz, Oguzhan, Ashok Regmi, John M. Anderies, and Armando A. Rodriguez. 2010. "Robustness, Vulnerability, and Adaptive Capacity in Small-Scale Socialecological Systems: The Pumpa Irrigation System in Nepal." *Ecology and Society* 15 (3): 39–69.

Cinner, Joshua E., Tim R. McClanahan, M. A. MacNeil, Nicholas A. Graham, Tim M. Daw, Ahmad Mukminin, David A. Feary et al. 2012. "Comanagement of Coral Reef Social-Ecological Systems." *Proceedings of the National Academy of Sciences of the United States of America* 109 (14): 5219–22. https://doi.org/10.1073/pnas.1121215109.

Clarke, Bertrand S., Ernest Fokoué, and Hao H. Zhang. 2009. *Principles and Theory for Data Mining and Machine Learning.* New York: Springer.

Cleaver, Frances. 2000. "Moral Ecological Rationality, Institutions and the Management of Common Property Resources." *Development and Change* 31 (2): 361–83.

Clutton-Brock, Tim H. 2002. "Breeding Together: Kin Selection and Mutualism in Cooperative Vertebrates." *Science* 296 (5565): 69–72. https://doi.org/10.1126/science.296.5565.69.

Clutton-Brock, Tim H., and G A. Parker. 1995. "Punishment in Animal Societies." *Nature* 373 (6511): 209–16. https://doi.org/10.1038/373209a0.

Coase, Ronald H. 1937. "The Nature of the Firm." *Economica* 4 (16): 386–405.

Coase, Ronald H. 1960. "The Problem of Social Cost." *Journal of Law and Economics* 3: 1–44.

Costanza, Robert. 1998. "Principles for Sustainable Governance of the Oceans." *Science* 281 (5374): 198–99. https://doi.org/10.1126/science.281.5374.198.

Cox, Michael. 2010. *Exploring the Dynamics of Social-Ecological Systems: The Case of the Taos Valley Acequias.* Bloomington, Indiana: Dissertation.

Cox, Michael, Gwen Arnold, and Sergio Villamayor Tomas. 2010. "A Review of Design Principles for Community-Based Natural Resource Management." *Ecology and Society* 15 (4): 38–57.

Cox, Michael, Sergio Villamayor-Tomas, Graham Epstein, Louisa Evans, Natalie C. Ban, Forrest Fleischman, Mateja Nenadovic, and Gustavo Garcia-Lopez. 2016. "Synthesizing Theories of Natural Resource Management and Governance." *Global Environmental Change* 39: 45–56. https://doi.org/10.1016/j.gloenvcha.2016.04.011.

Crawford, Sue, and Elinor Ostrom. 1995. "A Grammar of Institutions." *American Political Science Review* 89 (3): 582. https://doi.org/10.2307/2082975.

Creel, Scott, and Nancy M. Creel. 1995. "Communal Hunting and Pack Size in African Wild Dogs, Lycaon Pictus." *Animal Behaviour* 50: 1325–39.

Crespi, Bernard J. 2001. "The Evolution of Social Behavior in Microorganisms." *Trends in Ecology & Evolution* 16 (4): 178–83.

Dall, Sasha R. X., and Nina Wedell. 2005. "Evolutionary Conflict: Sperm Wars, Phantom Inseminations." *Current Biology* 15 (19): R801-R803.

Darwin, Charles. 1874. *Die Abstammung des Menschen.* Stuttgart: Kröner.

Dawes, Robyn M., Jeanne McTavish, and Harriet Shaklee. 1977. "Behavior, Communication, and Assumptions About Other People's Behavior in a Commons Dilemma Situation." *Journal of Personality and Social Psychology* 35 (1): 1–11.

Dawkins, Richard. 1976. *The Selfish Gene.* Oxford: Oxford University Press.

Dayton-Johnson, Jeff. 2000. "Determinants of Collective Action on the Local Commons: A Model with Evidence from Mexico." *Journal of Development Economics* 62: 181–208.

de Waal, Frans B. M. 1989. "Food Sharing and Reciprocal Obligations among Chimpanzees." *Journal of Human Evolution* 18 (5): 433–59.

Denant-Boemont, Laurent, David Masclet, and Charles N. Noussair. 2007. "Punishment, Counterpunishment and Sanction Enforcement in a Social Dilemma Experiment." *Economic Theory* 33: 145–67.

Diaz, Sandra, Josef Settele, Eduardo Brondizio, and et. al. 2019. "Summary for Policymakers of the Global Assessment Report on Biodiversity and Ecosystem Services of the Intergovernmental Science-Policy Platform on Biodiversity and Ecosystem Services." https://uwe-repository.worktribe.com/output/1493508/summary-for-policymakers-of-the-global-assessment-report-on-biodiversity-and-ecosystem-services-of-the-intergovernmental-science-policy-platform-on-biodiversity-and-ecosystem-services.

Diener, Ed, Ed Sandvik, Larry Seidlitz, and Marissa Diener. 1993. "The Relationship Between Income and Subjective Well-Being: Relative or Absolute?" *Social Indicators Research* 28: 195–223.

Dietz, Thomas, Nives Dolšak, Elinor Ostrom, and Paul C. Stern. 2002. "The Drama of the Commons." In Ostrom, Dietz, Dolšak, Stern, Stonich, and Weber 2002, 1–36.

Dobson, Andrew, David Lodge, Jackie Alder, Graeme S. Cumming, Juan Keymer, Jacquie McGlade, Hal Mooney et al. 2006. "Habitat Loss, Trophic Collapse, and the Decline of Ecosystem Services." *Ecology* 87 (8): 1915–24. https://doi.org/10.1890/0012-9658(2006)87[1915:HLTCAT]2.0.CO;2.

Doebeli, Michael, and Christoph Hauert. 2005. "Models of Cooperation Based on the Prisoner's Dilemma and the Snowdrift Game." *Ecology Letters* 8: 748–66.

Donlan, Rodney, and J W. Costerton. 2002. "Biofilms: Survival Mechanisms of Clinically Relevant Microorganisms." *Clinical Microbiological Reviews* 15 (2): 167–93.

Drea, Christina, and Alissa N. Carter. 2009. "Cooperative Problem Solving in a Social Carnivore." *Animal Behaviour* 78 (4): 967–77.

Edenhofer, Ottmar, Ramon Pichs-Madruga, Youba Sokona, Kristin Seyboth, Patrick Matschoss, Susanne Kadner, Timm Zwickel et al., eds. 2011. *IPCC Special Report on Renewable Energy Sources and Climate Change Mitigation.* Cambridge: Cambridge University Press.

Egas, Martijn, and Arno Riedl. 2008. "The Economics of Altruistic Punishment and the Maintenance of Cooperation." *Proceedings of the Royal Society B: Biological sciences* 275: 871–78. https://doi.org/10.1098/rspb.2007.1558.

Epstein, Graham, Jessica M. Vogt, Sarah K. Mincey, Michael Cox, and Burney Fischer. 2013. "Missing Ecology: Integrating Ecological Perspectives with the Social-Ecological System Framework." *International Journal of the Commons* 7 (2): 432–453. https://doi.org/10.18352/ijc.371.

Evans, Louisa, Nia Cherrett, and Diemuth Pemsl. 2011. "Assessing the Impact of Fisheries Co-Management Interventions in Developing Countries: A Meta-Analysis." *Journal of Environmental Management* 92 (8): 1938–49. https://doi.org/10.1016/j.jenvman.2011.03.010.

Falk, Armin, Ernst Fehr, and Urs Fischbacher. 2003. "On the Nature of Fair Behavior." *Economic Inquiry* 41 (1): 20–26.

Falk, Armin, and Urs Fischbacher. 2006. "A Theory of Reciprocity." *Games and Economic Behavior* 54: 293–315.

FAO. 2015. "Global Forest Resources Assessment 2015." Accessed September 22, 2015. http://www.fao.org/forest-resources-assessment/en/.

Feeny, David H. 1992. "Where Do We Go from Here? Implications for the Research Agenda." In Bromley, Feeny, Peters, Gilles, Oakerson, Runge, and Thomson, eds. 1992, 267–92.

Fehr, Ernst, Helen Bernhard, and Bettina Rockenbach. 2008. "Egalitarianism in Young Children." *Nature* 454 (7208): 1079–84.

Fehr, Ernst, and Urs Fischbacher. 2003. "The Nature of Human Altruism." *Nature* 425 (6960): 785–91. https://doi.org/10.1038/nature02043.

Fehr, Ernst, Urs Fischbacher, and Simon Gächter. 2002. "Strong Reciprocity, Human Cooperation, and the Enforcement of Social Norms." *Human Nature* 13 (1): 1–25. https://doi.org/10.1007/s12110-002-1012-7.

Fehr, Ernst, Urs Fischbacher, Bernhard von Rosenbladt, Jürgen Schupp, and Gert G. Wagner. 2003. "A Nation-Wide-Laboratory Examining Trust and Trustworthiness by Integrating Behavioral Experiments into Representative Surveys." Accessed July 8, 2012. http://www.cepr.org/pubs/dps/DP3858.asp.

Fehr, Ernst, and Simon Gächter. 2000. "Cooperation and Punishment in Public Goods Experiments." *American Economic Review* 90 (4): 980–94.

Fehr, Ernst, and Klaus M. Schmidt. 1999. "A Theory of Fairness, Competition, and Cooperation." *Quarterly Journal of Economics* 114 (3): 817–68.

Fischbacher, Urs, Simon Gächter, and Ernst Fehr. 2001. "Are People Conditionally Cooperative? Evidence from a Public Goods Experiment." *Economics Letters* 71 (3): 397–404.

Folke, Carl. 2006. "Resilience: The Emergence of a Perspective for Social–ecological Systems Analyses." *Global Environmental Change* 16 (3): 253–67. https://doi.org/10.1016/j.gloenvcha.2006.04.002.

Folke, Carl, Steve Carpenter, Thomas Elmqvist, Lance Gunderson, C S. Holling, and Brian Walker. 2002. "Resilience and Sustainable Development: Building Adaptive Capacity in a World of Transformations." *AMBIO: A Journal of the Human Environment* 31 (5): 437–40. https://doi.org/10.1579/0044-7447-31.5.437.

Frank, Robert H. 1988. *Passions Within Reason: The Strategic Role of the Emotions.* 1st ed. New York: Norton.

Franzen, Axel, and Sonja Pointner. 2012. "Anonymity in the Dictator Game Revisited." *Journal of Economic Behavior and Organization* 81 (1): 74–81. https://doi.org/10.1016/j.jebo.2011.09.005.

Freelon, D. 2013. "ReCal OIR: Ordinal, Interval, and Ratio Intercoder Reliability as a Web Service." *International Journal of Internet Science* 8 (1): 10–16.

Frey, Bruno S., and Stephan Meier. 2004. "Pro-Social Behavior in a Natural Setting." *Journal of Economic Behavior and Organization* 54 (1): 65–88.

Frey, Ulrich J. 2007. *Der blinde Fleck—Kognitive Fehler in der Wissenschaft und ihre evolutionsbiologischen Grundlagen.* Heusenstamm: Ontos.

Frey, Ulrich J. 2017a. "A Synthesis of Key Factors for Sustainability in Social-Ecological Systems." *Sustainability Science* 12 (4): 507–19. https://doi.org/10.1007/s11625-016-0395-z.

Frey, Ulrich J. 2017b. "Cooperative Strategies Outside the Laboratory — Evidence from a Long-Term Large-N-Study in Five Countries." *Evolution and Human Behavior* 38 (1): 109–16. https://doi.org/10.1016/j.evolhumbehav.2016.07.006.

Frey, Ulrich J., and Michael Cox. 2015. "Building a Diagnostic Ontology of Social-Ecological Systems." *International Journal of the Commons* 9 (2): 595–618. https://doi.org/10.18352/ijc.505.

Frey, Ulrich J., and Hannes Rusch. 2012. "An Evolutionary Perspective on the Long-Term Efficiency of Costly Punishment." *Biology and Philosophy* 27 (6): 811–31. https://doi.org/10.1007/s10539-012-9327-1.

Frey, Ulrich J., and Hannes Rusch. 2013. "Using Artificial Neural Networks for the Analysis of Social-Ecological Systems." *Ecology and Society* 18 (2): 40. https://doi.org/10.5751/ES-05202-180240.

Frey, Ulrich J., Sergio Villamayor-Tomas, and Insa Theesfeld. 2016. "A Continuum of Governance Regimes: A New Perspective on Co-Management in Irrigation Systems." *Environmental Science & Policy* 66: 73–81. https://doi.org/10.1016/j.envsci.2016.08.008.

Gächter, Simon, Benedikt Herrmann, and Christian Thöni. 2004. "Trust, Voluntary Cooperation, and Socio-Economic Background: Survey and Experimental Evidence." *Journal of Economic Behavior and Organization* 55: 505–31.

Gächter, Simon, Elke Renner, and Martin Sefton. 2008. "The Long-Run Benefits of Punishment." *Science* 322: 1510. https://doi.org/10.1126/science.1164744.

Gentry, Alwyn H. 1988. "Changes in Plant Community Diversity and Floristic Composition on Environmental and Geographical Gradients." *Annals of the Missouri Botanical Garden* 75 (1): 1–34.

Gerber, Anke, and Philipp C. Wichardt. 2009. "Providing Public Goods in the Absence of Strong Institutions." *Journal of Public Economics* 93: 429–39.

Gevrey, Muriel, Ioannis Dimopoulos, and Sovan Lek. 2003. "Review and Comparison of Methods to Study the Contribution of Variables in Artificial Neural Network Models." *Ecological Modelling* 160 (3): 249–64. https://doi.org/10.1016/S0304-3800(02)00257-0.

Gibson, Clark C., John T. Williams, and Elinor Ostrom. 2005. "Local Enforcement and Better Forests." *World Development* 33 (2): 273–84.

Gilovich, Thomas, Dale Griffin, and Daniel Kahneman, eds. 2002. *Heuristics and Biases: The Psychology of Intuitive Judgment.* Cambridge: Cambridge University Press.

Gintis, Herbert. 2000. "Beyond Homo Economicus: Evidence from Experimental Economics." *Ecological Economics* 35: 311–22.

Glaser, Marion, Patrick Christie, Karen Diele, Larissa Dsikowitzky, Sebastian Ferse, Inga Nordhaus, Achim Schlüter, Kathleen Schwerdtner Mañez, and Christian Wild. 2012. "Measuring and Understanding Sustainability-Enhancing Processes in Tropical Coastal and Marine Social-Ecological Systems." *Current Opinion in Environmental Sustainability* 4 (3): 300–308. https://doi.org/10.1016/j.cosust.2012.05.004.

Gruber, Eva. 2008. "Key Principles of Community-Based Natural Resource Management: A Synthesis and Interpretation of Identified Effective Approaches for Managing the Commons." *Environmental Management* 45: 52–66.

Guala, Francesco. 2012. "Reciprocity: Weak or Strong? What Punishment Experiments Do (and Do Not) Demonstrate." *Behavioral and Brain Sciences* 35 (1): 1–15. https://doi.org/10.1017/S0140525X11000069.

Gürerk, Özgür, Bernd Irlenbusch, and Bettina Rockenbach. 2006. "The Competitive Advantage of Sanctioning Institutions." *Science* 312 (5770): 108–11. https://doi.org/10.1126/science.1123633.

Güth, Werner, Rolf Schmittberger, and Bernd Schwarze. 1982. "An Experimental Analysis of Ultimatum Bargaining." *Journal of Economic Behavior and Organization* 3 (4): 367–88. https://doi.org/10.1016/0167-2681(82)90011-7.

Gutiérrez, Nicolás L., Ray Hilborn, and Omar Defeo. 2011. "Leadership, Social Capital and Incentives Promote Successful Fisheries." *Nature* 470 (7334): 386–89. https://doi.org/10.1038/nature09689.

Hagedorn, Konrad. 2008. "Particular Requirements for Institutional Analysis in Nature-Related Sectors." *European Review of Agricultural Economics* 35 (3): 357–84. https://doi.org/10.1093/erae/jbn019.

Hagen, Edward H., and Peter Hammerstein. 2006. "Game Theory and Human Evolution: A Critique of Some Recent Interpretations of Experimental Games." *Theoretical population biology* 69 (3): 339–48. https://doi.org/10.1016/j.tpb.2005.09.005.

Hallmann, Caspar A., Martin Sorg, Eelke Jongejans, Henk Siepel, Nick Hofland, Heinz Schwan, Werner Stenmans et al. 2017. "More Than 75 Percent Decline over 27 Years in Total Flying Insect Biomass in Protected Areas." *PloS one* 12 (10): e0185809. https://doi.org/10.1371/journal.pone.0185809.

Hamilton, William D. 1964. "The Genetical Evolution of Social Behaviour I & II." *Journal of Theoretical Biology* 7: 1–52.

Hansen, Matthew C., Stephen V. Stehman, and Peter V. Potapov. 2010. "Quantification of Global Gross Forest Cover Loss." *Proceedings of the National Academy of Sciences of the United States of America* 107 (19): 8650–55. https://doi.org/10.1073/pnas.0912668107.

Harbaugh, William T., Kate Krause, and Stephan G. Liday. 2003. "Bargaining by Children." *Unpublished manuscript*: 1–40.

Hardin, Garrett. 1968. "The Tragedy of the Commons." *Science* 162: 1243–48.

Hardin, Garrett. 1993. *Living Within Limits: Ecology, Economics, and Population Taboos.* Oxford: Oxford University Press.

Hawkes, Kristen, and Rebecca B. Bird. 2002. "Showing Off, Handicap Signaling, and the Evolution of Men's Work." *Evolutionary Anthropology* 11 (2): 58–67.

He, Fengzhi, Christiane Zarfl, Vanessa Bremerich, Jonathan N. W. David, Zeb Hogan, Gregor Kalinkat, Klement Tockner, and Sonja C. Jähnig. 2019. "The Global Decline of Freshwater Megafauna." *Global change biology* 25 (11): 3883–92. https://doi.org/10.1111/gcb.14753.

Heinsohn, Robert, and Craig Packer. 1995. "Complex Cooperative Strategies in Group-Territorial African Lions." *Science* 269 (5228): 1260–62. https://doi.org/10.1126/science.7652573.

Helanterä, Heikki, and Liselotte Sundström. 2007. "Worker Policing and Nest Mate Recognition in the Ant Formica Fusca." *Behavioral Ecology and Sociobiology* 61 (8): 1143–49. https://doi.org/10.1007/s00265-006-0327-5.

Helbing, Dirk, and Wenjian Yu. 2009. "The Outbreak of Cooperation among Success-Driven Individuals Under Noisy Conditions." *Proceedings of the National Academy of Sciences of the United States of America* 106 (10): 3680–85.

Henrich, Joseph, Robert Boyd, Samuel Bowles, Colin F. Camerer, Ernst Fehr, Herbert Gintis, and Richard McElreath. 2001. "In Search of Homo Economicus: Behavioral Experiments in 15 Small-Scale Societies." *American Economic Review* 91 (2): 73–78.

Henrich, Joseph, Jean Ensminger, Richard McElreath, Abigail Barr, Clark H. Barrett, Alexander Bolyanatz, Juan-Camilo Cardenas et al. 2010. "Markets, Religion, Community Size, and the Evolution of Fairness and Punishment." *Science* 327 (5972): 1480–84.

Henrich, Joseph, Steven J. Heine, and Ara Norenzayan. 2010. "The Weirdest People in the World?" *Behavioral and Brain Sciences* 33: 61–135.

Herrfahrdt-Pähle, Elke, and Claudia Pahl-Wostl. 2012. "Continuity and Change in Social-Ecological Systems: The Role of Institutional Resilience." *Ecology and Society* 17 (2): 8. https://doi.org/10.5751/ES-04565-170208.

Herrmann, Benedikt, Christian Thöni, and Simon Gächter. 2008. "Antisocial Punishment Across Societies." *Science* 319: 1362–67. https://doi.org/10.1126/science.1153808.

Hess, Charlotte. 2008. "Mapping the New Commons." Accessed April 01, 2020. https://papers.ssrn.com/sol3/papers.cfm?abstract_id=1356835.

Hess, Charlotte, and Elinor Ostrom. 2003. "Ideas, Artifacts, and Facilities: Information as a Common-Pool Resource." *Law and Contemporary Problems* 66: 111–46.

Hileman, Jacob, Paul Hicks, and Richard Jones. 2015. "An Alternative Framework for Analysing and Managing Conflicts in Integrated Water Resources Management (IWRM): Linking Theory and Practice." *International Journal of Water Resources Development* 32 (5): 675–91. https://doi.org/10.1080/07900627.2015.1076719.

Hinkel, Jochen, Michael Cox, Maja Schlüter, Claudia R. Binder, and Thomas Falk. 2015. "A Diagnostic Procedure for Applying the Social-Ecological Systems Framework in Diverse Cases." *Ecology and Society* 20 (1): 32. https://doi.org/10.5751/ES-07023-200132.

Hirota, Marina, Milena Holmgren, Egbert Van Nes, and Marten Scheffer. 2011. "Global Resilience of Tropical Forest and Savanna to Critical Transitions." *Science* 334 (6053): 232–35. https://doi.org/10.1126/science.1210657.

Holling, C S. 1973. "Resilience and Stability of Ecological Systems." *Annual Review of Ecology and Systematics* 4: 1–23. http://www.jstor.org/stable/2096802.

Hsee, Christopher K., Yang Yang, Naihe Li, and Luxi Shen. 2009. "Wealth, Warmth, and Well-Being: Whether Happiness Is Relative or Absolute Depends on Whether It Is about Money, Acquisition, or Consumption." *Journal of Marketing Research* 46 (3): 396–409.

Huitema, Dave, Erik Mostert, Wouter Egas, Sabine Moellenkamp, Claudia Pahl-Wostl, and Resul Yalcin. 2009. "Adaptive Water Governance: Assessing the Institutional Prescriptions of Adaptive (Co-)Management from a Governance Perspective and Defining a Research Agenda." *Ecology and Society* 14 (1): 26. http://www.ecologyandsociety.org/vol14/iss1/art26/.

IPCC. 2018. "Global Warming of 1.5°C. An IPCC Special Report on the Impacts of Global Warming of 1.5°C Above Pre-Industrial Levels and Related Global Greenhouse Gas Emission Pathways, in the Context of Strengthening the Global Response to the Threat of Climate Change, Sustainable Development, and Efforts to Eradicate Poverty." https://www.ipcc.ch/sr15/.

Isaac, R M., K F. McCue, and Charles R. Plott. 1985. "Public Goods Provision in an Experimental Environment." *Journal of Public Economics* 26: 51–74.

Isaac, R M., James M. Walker, and Arlington W. Williams. 1994. "Group Size and the Voluntary Provision of Public Goods: Experimental Evidence Utilizing Large Groups." *Journal of Public Economics* 54 (1): 1–36.

Jandér, K C., and Edward A. Herre. 2010. "Host Sanctions and Pollinator Cheating in the Fig Tree-Fig Wasp Mutualism." *Proceedings of the Royal Society B: Biological sciences* 277 (1687): 1481–88. https://doi.org/10.1098/rspb.2009.2157.

Jentoft, Svein, Bonnie J. McCay, and Douglas C. Wilson. 1998. "Social Theory and Fisheries Co-Management." *Marine Policy* 22 (4-5): 423–36. https://doi.org/10.1016/S0308-597X(97)00040-7.

Jentoft, Svein, Thijs C. van Son, and Maiken Bjørkan. 2007. "Marine Protected Areas: A Governance System Analysis." *Human Ecology* 35 (5): 611–22. https://doi.org/10.1007/s10745-007-9125-6.

Johnstone, Rufus A. 1995. "Sexual Selection, Honest Advertisement and the Handicap Principle: Reviewing the Evidence." *Biological Review* 70: 1–65.

Joshi, Neeraj N. 2000. "Institutional Opportunities and Constraints in the Performance of Farmer-Managed Irrigation Systems in Nepal." *Asia-Pacific journal of rural development.* 10 (2): 67–92.

Karlan, Dean S. 2005. "Using Experimental Economics to Measure Social Capital and Predict Financial Decisions." *American Economic Review* 95 (5): 1688–99.

Kerr, Benjamin, Claudia Neuhauser, Brendan J. M. Bohannan, and Antony M. Dean. 2006. "Local Migration Promotes Competitive Restraint in a Host-Pathogen 'Tragedy of the Commons'." *Nature* 442: 75–78.

Khan, Javet, Jun S. Wei, Markus Ringnér, Lao H. Saal, Marc Ladanyi, Frank Westermann, Frank Berthold et al. 2001. "Classification and Diagnostic Prediction of Cancers Using Gene Expression Profiling and Artificial Neural Networks." *Nature Medicine* 7 (6): 673–79.

Kiers, E T., Robert A. Rousseau, Stuart A. West, and R F. Denison. 2003. "Host Sanctions and the Legume-Rhizobium Mutualism." *Nature* 425: 78–81.

Klein, Peter G. 1998. "New Institutional Economics." *SSRN Electronic Journal.* https://doi.org/10.2139/ssrn.115811.

Knutti, Reto, Thomas F. Stocker, Fortunat Joos, and G-K Plattner. 2003. "Probabilistic Climate Change Projections Using Neural Networks." *Climate Dynamics* 21: 257–72.

Kocher, Martin G., Todd L. Cherry, Stephan Kroll, Robert J. Netzer, and Matthias Sutter. 2008. "Conditional Cooperation on Three Continents." *Economics Letters* 101: 175–78.

Koopmans, Ruud, and Susanne Rebers. 2009. "Collective Action in Culturally Similar and Dissimilar Groups: An Experiment on Parochialism, Conditional Cooperation, and Their Linkages." *Evolution and Human Behavior* 30 (3): 201–11.

Kopfmüller, Jürgen. 2001. *Nachhaltige Entwicklung integrativ betrachtet: Konstitutive Elemente, Regeln, Indikatoren.* Global zukunftsfähige Entwicklung - Perspektiven für Deutschland 1. Berlin: Ed. Sigma.

Kosfeld, Michael, Akira Okada, and Arno Riedl. 2009. "Institution Formation in Public Goods Games." *American Economic Review* 99 (4): 1335–55.

Krebs, John R., and Nick B. Davies, eds. 1978. *Behavioral Ecology: An Evolutionary Approach.* Oxford: Blackwell.

Kreft, Jan-Ulrich. 2004. "Biofilms Promote Altruism." *Microbiology* 150: 2751–60.

Krippendorff, Klaus. 2004. "Reliability in Content Analysis." *Human Communication Research* 30 (3): 411–33. https://doi.org/10.1111/j.1468-2958.2004.tb00738.x.

Kühberger, Anton. 1998. "The Influence of Framing on Risky Decisions: A Meta-Analysis." *Organizational Behaviour and Human Decision Processes* 75: 23–55.

Kurzban, Robert. 2001. "The Social Psychophysics of Cooperation: Nonverbal Communication in a Public Goods Game." *Journal of Nonverbal Behavior* 25 (4): 241–59.

Lam, Marco A., and Elinor Ostrom. 2010. "Analyzing the Dynamic Complexity of Development Interventions: Lessons from an Irrigation Experiment in Nepal." *Policy Sciences* 43: 1–25.

Lam, Wai Fung. 1996. "Improving the Performance of Small-Scale Irrigation Systems: The Effects of Technological Investments and Governance Structure on Irrigation Performance in Nepal." *World Development* 24 (8): 1301–15.

Lam, Wai Fung. 1998. *Governing Irrigation Systems in Nepal: Institutions, Infrastructure, and Collective Action.* San Francisco: Institute for Contemporary Studies.

Lammerts van Bueren, Erik M., and Esther M. Blom. 1997. *Hierarchical Framework for the Formulation of Sustainable Forest Management Standards*. Leiden: Trobenbos Foundation.

Laury, Susan K., James M. Walker, and Arlington W. Williams. 1995. "Anonymity and the Voluntary Provision of Public Goods." *Journal of Economic Behavior and Organization* 27: 365–80.

LeCun, Yann, Yoshua Bengio, and Geoffrey Hinton. 2015. "Deep Learning." *Nature* 521 (7553): 436–44. https://doi.org/10.1038/nature14539.

Ledyard, John O. 1995. "Public Goods: A Survey of Experimental Research." In *The Handbook of Experimental Economics*, edited by John H. Kagel and Alvin E. Roth, 111–94. Princeton: Princeton University Press.

Lehmann, Laurent, and Laurent Keller. 2006. "The Evolution of Cooperation and Altruism—a General Framework and a Classification of Models." *Journal of Evolutionary Biology* 19: 1365–76.

Leigh, Jr E G. 2010. "The Evolution of Mutualism." *Journal of Evolutionary Biology* 23 (12): 2507–28. https://doi.org/10.1111/j.1420-9101.2010.02114.x.

Leimar, Olof, and Peter Hammerstein. 2001. "Evolution of Cooperation through Indirect Reciprocity." *Proceedings of the Royal Society B: Biological sciences* 268: 745–53. https://doi.org/10.1098/rspb.2000.1573.

Leslie, Heather M., Xavier Basurto, Mateja Nenadovic, Leila Sievanen, Kyle C. Cavanaugh, Juan J. Cota-Nieto, Brad E. Erisman et al. 2015. "Operationalizing the Social-Ecological Systems Framework to Assess Sustainability." *Proceedings of the National Academy of Sciences of the United States of America* 112 (19): 5979–84. https://doi.org/10.1073/pnas.1414640112.

Levitt, Steven D., and John A. List. 2007. "What Do Laboratory Experiments Measuring Social Preferences Reveal about the Real World?" *Journal of Economic Perspectives* 21 (2): 153–74. https://doi.org/10.2307/30033722.

Limburg, Karin E., Robert V. O'Neill, Robert Costanza, and Stephen Farber. 2002. "Complex Systems and Valuation." *Ecological Economics* 41 (3): 409–20. https://doi.org/10.1016/S0921-8009(02)00090-3.

List, John A. 2006. "The Behavioralist Meets the Market: Measuring Social Preferences and Reputation Effects in Actual Transactions." *Journal of Political Economy* 114 (1): 1–37. https://doi.org/10.1086/498587.

List, John A., and Michael K. Price. 2009. "The Role of Social Connections in Charitable Fundraising: Evidence from a Natural Field Experiment." *Journal of Economic Behavior and Organization* 69: 160–69.

Lombard, Matthew, Jennifer Snyder-Duch, and Cheryl C. Bracken. 2002. "Content Analysis in Mass Communication: Assessment and Reporting of Intercoder Reliability." *Human Communication Research* 28 (4): 587–604. https://doi.org/10.1111/j.1468-2958.2002.tb00826.x.

MacDougall, Andrew S., Kevin S. McCann, Gabriel Gellner, and Roy Turkington. 2013. "Diversity Loss with Persistent Human Disturbance Increases Vulnerability to Ecosystem Collapse." *Nature* 494 (7435): 86–89. https://doi.org/10.1038/nature11869.

Madrigal, Róger, Francisco Alpízar, and Achim Schlüter. 2011. "Determinants of Performance of Community-Based Drinking Water Organizations." *World Development* 39 (9): 1663–75. https://doi.org/10.1016/j.worlddev.2011.02.011.

Mares, Rafael, Andrew Young, and Tim H. Clutton-Brock. 2012. "Individual Contributions to Territory Defence in a Cooperative Breeder: Weighing up the Benefits and Costs." *Proceedings of the Royal Society B: Biological sciences* 279 (1744): 3989–95. https://doi.org/10.1098/rspb.2012.1071.

Marlowe, Frank W. 2009. "Hadza Cooperation Second-Party Punishment, Yes, Third-Party Punishment, No." *Human Nature* 20: 417–30.

Marwell, Gerald, and Ruth E. Ames. 1981. "Economists Free Ride, Does Anyone Else? Experiments on the Provision of Public Goods, IV." *Journal of Public Economics* 15 (3): 295–310.

May, Robert M. 1999. "Unanswered Questions in Ecology." *Philosophical Transactions of The Royal Society of London: Biological Sciences* 354 (1392): 1951–59. https://doi.org/10.1098/rstb.1999.0534.

McGinnis, Michael, and Elinor Ostrom. 2014. "Social-Ecological System Framework: Initial Changes and Continuing Challenges." *Ecology and Society* 19 (2): 30–42. https://doi.org/10.5751/ES-06387-190230.

McKean, Margaret A. 1992. "Management of Traditional Common Lands (Iriaichi) in Japan." In Bromley, Feeny, Peters, Gilles, Oakerson, Runge, and Thomson, eds. 1992, 63–98.

Meadows, Donella H., Dennis L. Meadows, Jorgen Randers, and William W. Behrens. 1972. *The Limits to Growth: A Report for the Club of Rome's Project on the Predicament of Mankind.* New York: Universe Books.

Mehrotra, Kishan, Chilukuri K. Mohan, and Sanjay Ranka. 1997. *Elements of Artificial Neural Networks.* Cambridge, MA: MIT Press.

Meinzen-Dick, Ruth. 2007. "Beyond Panaceas in Water Institutions." *Proceedings of the National Academy of Sciences of the United States of America* 104 (39): 15200–15205.

Ménard, Claude, and Mary M. Shirley, eds. 2008. *Handbook of New Institutional Economics.* Berlin: Springer.

Milinski, Manfred, Dirk Semmann, Hans-Jürgen Krambeck, and Jochen Marotzke. 2006. "Stabilizing the Earth's Climate Is Not a Losing Game: Supporting Evidence from Public Goods Experiments." *Proceedings of the National Academy of Sciences of the United States of America* 103 (11): 3994–98.

Mitchell, Gregory. 2012. "Revisiting Truth or Triviality: The External Validity of Research in the Psychological Laboratory." *Perspectives on Psychological Science* 7 (2): 109–17. https://doi.org/10.1177/1745691611432343.

Mitchell, Tom M. 1997. *Machine Learning.* New York: McGraw-Hill.

Mulder, Raoul A., and Naomi E. Langmore. 1993. "Dominant Males Punish Helpers for Temporary Defection in Superb Fairy-Wrens." *Animal Behaviour* 45: 830–33.

Myatt, Glenn J., and Wayne P. Johnson. 2009. *A Practical Guide to Data Visualization, Advanced Data Mining Methods, and Applications.* Hoboken, NJ: Wiley.

Myers, Ransom A., and Boris Worm. 2003. "Rapid Worldwide Depletion of Predatory Fish Communities." *Nature* 423 (6937): 280–83. https://doi.org/10.1038/nature01610.

Nagendra, Harini. 2007. "Drivers of Reforestation in Human-Dominated Forests." *Proceedings of the National Academy of Sciences of the United States of America* 104 (39): 15218–23.

Nash, John. 1951. "Non-Cooperative Games." *The Annals of Mathematics* 54 (2): 286–95.

Neuhauser, Claudia, and Joseph E. Fargione. 2004. "A Mutualism-Parasitism Continuum Model and Its Application to Plant-Mycorrhizae Interactions." *Ecological Modelling* 177 (3-4): 337–52. https://doi.org/10.1016/j.ecolmodel.2004.02.010.

Newell, Allen, and Herbert A. Simon. 1972. *Human Problem Solving.* New Jersey: Prentice Hall.

Nikiforakis, Nikos, and Hans-Theo Normann. 2008. "A Comparative Statics Analysis of Punishment in Public-Good Experiments." *Experimental Economics* 11 (4): 358–69. https://doi.org/10.1007/s10683-007-9171-3.

Nilsson, Tobias. 2001. "Management of Communal Grazing Land: A Case Study on Institutions for Collective Action in Endabeg Village, Tanzania." Accessed February 05, 2009. http://dlc.dlib.indiana.edu/archive/00000961/00/WebThesis.pdf.

Noë, Ronald. 2006. "Cooperation Experiments: Coordination through Communication versus Acting Apart Together." *Animal Behaviour* 71 (1): 1–18. https://doi.org/10.1016/j.anbehav.2005.03.037.

Noë, Ronald, Jan A. R. A. M. van Hooff, and Peter Hammerstein. 2001. *Economics in Nature: Social Dilemmas and Biological Markets.* New York: Cambridge University Press.

Nowak, Martin A. 2006. "Five Rules for the Evolution of Cooperation." *Science* 314: 1560–63.

Nowak, Martin A., and Robert M. May. 1992. "Evolutionary Games and Spatial Chaos." *Nature* 359: 826–29.

Nowak, Martin A., and Karl Sigmund. 1998. "Evolution of Indirect Reciprocity by Image Scoring." *Nature* 393: 573–77.

Nowell, Clifford, and Sarah Tinkler. 1994. "The Influence of Gender on the Provision of a Public Good." *Journal of Economic Behavior and Organization* 25: 25–36.

Oakerson, Ronald J. 1992. "Analyzing the Commons: A Framework." In Bromley, Feeny, Peters, Gilles, Oakerson, Runge, and Thomson, eds. 1992, 41–59.

Ockenfels, Axel, and Joachim Weimann. 1999. "Types and Patterns: An Experimental East-West-German Comparison of Cooperation and Solidarity." *Journal of Public Economics* 71: 275–87.

OECD. 1994. *Environmental Indicators: OECD Core Set.* Paris: Organisation for Economic Co-operation and Development.

Olden, Julian D., Michael K. Joy, and Russell G. Death. 2004. "An Accurate Comparison of Methods for Quantifying Variable Importance in Artificial Neural Networks Using Simulated Data." *Ecological Modelling* 178 (3-4): 389–97. https://doi.org/10.1016/j.ecolmodel.2004.03.013.

Olson, Mancur. 1968. *Die Logik des kollektiven Handelns: Kollektivgüter und die Theorie der Gruppen.* Tübingen: Mohr Siebeck.

Olsson, Per, Carl Folke, and Fikret Berkes. 2004. "Adaptive Comanagement for Building Resilience in Social-Ecological Systems." *Environmental Management* 34 (1): 75–90.

O'Neill, Michael C., and Li Song. 2003. "Neural Network Analysis of Lymphoma Microarray Data: Prognosis and Diagnosis Near-Perfect." *BMC Bioinformatics* 4 (13): 1–12.

Ones, Umut, and Louis Putterman. 2007. "The Ecology of Collective Action: A Public Goods and Sanctions Experiment with Controlled Group Formation." *Journal of Economic Behavior and Organization* 62 (4): 495–521.

Oosterbeek, Hessel, Randolf Sloof, and Gijs van de Kuilen. 2004. "Cultural Differences in Ultimatum Game Experiments: Evidence from a Meta-Analysis." *Experimental Economics* 7 (2): 171–88.

Ostrom, Elinor. 1990. *Governing the Commons: The Evolution of Institutions for Collective Action.* Cambridge: Cambridge University Press.

Ostrom, Elinor. 1992a. *Crafting Institutions for Self-Governing Irrigation Systems.* San Francisco: Institute for Contemporary Studies.

Ostrom, Elinor. 1992b. "The Rudiments of a Theory of the Origins, Survival, and Performance of Common-Property Institutions." In Bromley, Feeny, Peters, Gilles, Oakerson, Runge, and Thomson, eds. 1992, 293–318.

Ostrom, Elinor. 1998. "A Behavioral Approach to the Rational Choice Theory of Collective Action." *American Political Science Review* 92 (1): 1–22.

Ostrom, Elinor. 2005. *Understanding Institutional Diversity*. Princeton: Princeton University Press.

Ostrom, Elinor. 2007. "A Diagnostic Approach for Going Beyond Panaceas." *Proceedings of the National Academy of Sciences of the United States of America* 104 (39): 15181–87.

Ostrom, Elinor. 2008. "Frameworks and Theories of Environmental Change." *Global Environmental Change* 18 (2): 249–52. https://doi.org/10.1016/j.gloenvcha.2008.01.001.

Ostrom, Elinor. 2009. "A General Framework for Analyzing Sustainability of Social-Ecological Systems." *Science* 325: 419–22.

Ostrom, Elinor, and Michael Cox. 2010. "Moving Beyond Panaceas: A Multi-Tiered Diagnostic Approach for Social-Ecological Analysis." *Environmental Conservation* 37 (04): 451–63. https://doi.org/10.1017/S0376892910000834.

Ostrom, Elinor, Thomas Dietz, Nives Dolšak, Paul C. Stern, Susan Stonich, and Elke U. Weber, eds. 2002. *The Drama of the Commons*. Washington: National Academy Press.

Ostrom, Elinor, Roy Gardner, and James M. Walker. 1994. *Rules, Games, and Common-Pool Resources*. Ann Arbor: University of Michigan Press.

Ostrom, Elinor, Marco A. Janssen, and John M. Anderies. 2007. "Going Beyond Panaceas." *Proceedings of the National Academy of Sciences of the United States of America* 104 (39): 15176–78.

Ostrom, Elinor, Wai Fung Lam, Prachanda Pradhan, and Ganesh P. Shivakoti. 2011. *Improving Irrigation in Asia: Sustainable Performance of an Innovative Intervention in Nepal*. Cheltenham: Edward Elgar.

Ostrom, Elinor, James M. Walker, and Roy Gardner. 1992. "Covenants with and Without a Sword: Self-Governance Is Possible." *American Political Science Review* 86 (2): 404–17.

Pagdee, Adcharaporn, Yeon-Su Kim, and P J. Daugherty. 2006. "What Makes Community Forest Management Successful: A Meta-Study from Community Forests throughout the World." *Society & Natural Resources* 19: 33–52.

Page, Talbot, Louis Putterman, and Bulent Unel. 2005. "Voluntary Association in Public Goods Experiments: Reciprocity, Mimicry and Efficiency." *Economic Journal* 115 (506): 1032–53. https://doi.org/10.1111/j.1468-0297.2005.01031.x.

Pahl-Wostl, Claudia. 2009. "A Conceptual Framework for Analysing Adaptive Capacity and Multi-Level Learning Processes in Resource Governance Regimes." *Global Environmental Change* 19 (3): 354–65. https://doi.org/10.1016/j.gloenvcha.2009.06.001.

Pahl-Wostl, Claudia, Marc Craps, Art Dewulf, Erik Mostert, David Tabara, and Tharsi taillieu. 2007. "Social Learning and Water Resources Management." *Ecology and Society* 12 (2): 1–19. http://www.ecologyandsociety.org/vol12/iss2/art5/.

Pahl-Wostl, Claudia, Carlo Giupponi, Keith Richards, Claudia R. Binder, Alex de Sherbinin, Detlef Sprinz, Theo Toonen, and Caroline van Bers. 2013. "Transition Towards a New Global Change Science: Requirements for Methodologies, Methods, Data and Knowledge." *Environmental Science & Technology* 28: 36–47. https://doi.org/10.1016/j.envsci.2012.11.009.

Pahl-Wostl, Claudia, Georg Holtz, Britta Kastens, and Christian Knieper. 2010. "Analyzing Complex Water Governance Regimes: The Management and Transition

Framework." *Environmental Science & Technology* 13 (7): 571–81. https://doi.org/10.1016/j.envsci.2010.08.006.

Palacios-Huerta, Ignacio, and Oscar Volij. 2006. "Field Centipedes." Accessed October 29, 2008. http://home.cerge-ei.cz/ortmann/trentocourse/Palacios_Huerta_Volij_Field_Centepedes.pdf.

Palfrey, Thomas R., and Jeffrey E. Prisbrey. 1997. "Anomalous Behavior in Public Goods Experiments: How Much and Why?" *American Economic Review* 87 (5): 829–46.

Partelow, Stefan. 2015. "Coevolving Ostrom's Social-Ecological Systems (SES) Framework and Sustainability Science: Four Key Co-Benefits." *Sustainability Science* 11 (3): 399–410. https://doi.org/10.1007/s11625-015-0351-3.

Partelow, Stefan. 2018. "A Review of the Social-Ecological Systems Framework: Applications, Methods, Modifications, and Challenges." *Ecology and Society* 23 (4): 36. https://doi.org/10.5751/ES-10594-230436.

Peng, Changhui, and Xuezhi Wen. 1999. "Recent Applications of Artificial Neural Networks in Forest Resource Management: An Overview." Accessed September 28, 2019. https://www.aaai.org/Papers/Workshops/1999/WS-99-07/WS99-07-003.pdf.

Peters, H E., A S. Ünür, Jeremy Clark, and William D. Schulze. 2004. "Free-Riding and the Provision of Public Goods in the Family: A Laboratory Experiment." *International Economic Review* 45 (1): 283–99.

Plummer, Ryan, and Derek R. Armitage. 2007. "A Resilience-Based Framework for Evaluating Adaptive Co-Management: Linking Ecology, Economics and Society in a Complex World." *Ecological Economics* 61 (1): 62–74. https://doi.org/10.1016/j.ecolecon.2006.09.025.

Pomeroy, Robert S., Brenda M. Katon, and Ingvild Harkes. 1998. "Fisheries Co-Management: Key Conditions and Principles Drawn from Asian Experiences." *Paper presented at the 7th annual conference of the IASCP*: 1–23. Accessed February 28, 2009.

Pomeroy, Robert S., Brenda M. Katon, and Ingvild Harkes. 2001. "Conditions Affecting the Success of Fisheries Co-Management: Lessons from Asia." *Marine Policy* 25 (3): 197–208. https://doi.org/10.1016/S0308-597X(01)00010-0.

Poteete, Amy R., Marco A. Janssen, and Elinor Ostrom. 2010. *Working Together: Collective Action, the Commons, and Multiple Methods in Practice.* Princeton: Princeton University Press.

Poteete, Amy R., and Elinor Ostrom. 2008. "Fifteen Years of Empirical Research on Collective Action in Natural Resource Management: Struggling to Build Large-N Databases Based on Qualitative Research." *World Development* 36 (1): 176–95.

Pretty, Jules. 2003. "Social Capital and the Collective Management of Resources." *Science* 302 (5652): 1912–14. https://doi.org/10.1126/science.1090847.

Price, Michael E. 2006. "Monitoring, Reputation, and 'Greenbeard' Reciprocity in a Shuar Work Team." *Journal of Organizational Behavior* 27: 201–19.

Raihani, Nichola J., Alex Thornton, and Redouan Bshary. 2012. "Punishment and Cooperation in Nature." *Trends in Ecology & Evolution* 27 (5): 288–95. https://doi.org/10.1016/j.tree.2011.12.004.

Rainey, Paul B., and Katrina Rainey. 2003. "Evolution of Cooperation and Conflict in Experimental Bacterial Populations." *Nature* 425 (6953): 72–74.

Rankin, Daniel J. 2011. "The Social Side of Homo Economicus." *Trends in Ecology & Evolution* 26 (1): 1–3. https://doi.org/10.1016/j.tree.2010.10.005.

Rankin, Daniel J., Katja Bargum, and Hanna Kokko. 2007. "The Tragedy of the Commons in Evolutionary Biology." *Trends in Ecology & Evolution* 22 (12): 643–51.

Ratnieks, Francis L., Kevin R. Foster, and Tom Wenseleers. 2006. "Conflict Resolution in Insect Societies." *Annual Review of Entomology* 51: 581–608. https://doi.org/10.1146/annurev.ento.51.110104.151003.

Rawls, John. 1979. *Eine Theorie der Gerechtigkeit.* Frankfurt am Main: Suhrkamp.

Reed, Russell D., and Robert J. Marks. 1999. *Neural Smithing: Supervised Learning in Feedforward Artificial Neural Networks.* Cambridge, Massachusetts: MIT Press.

Reichhuber, Anke, Eva Camacho, and Till Requate. 2009. "A Framed Field Experiment on Collective Enforcement Mechanisms with Ethiopian Farmers." *Environment, Development and Sustainability* 14: 641–63.

Resnick, Paul, Richard J. Zeckhauser, John Swanson, and Kate Lockwood. 2006. "The Value of Reputation on EBay: A Controlled Experiment." *Experimental Economics* 9: 79–101.

Rilling, James K., David A. Gutman, Thorsten R. Zeh, Giuseppe Pagnoni, Gregory S. Berns, and Clinton D. Kilts. 2002. "A Neural Basis for Social Cooperation." *Neuron* 35: 395–405.

Rockenbach, Bettina, and Manfred Milinski. 2006. "The Efficient Interaction of Indirect Reciprocity and Costly Punishment." *Nature* 444: 718–23. https://doi.org/10.1038/nature05229.

Rova, Silvia, and Fabio Pranovi. 2017. "Analysis and Management of Multiple Ecosystem Services Within a Social-Ecological Context." *Ecological Indicators* 72: 436–43. https://doi.org/10.1016/j.ecolind.2016.07.050.

Rowley, Henry A., Shomeet Baluja, and Takeo Kanade. 1998. "Neural Network-Based Face Detection." *IEEE Transactions on PAMI* 20 (1): 23–28.

Russell, Andrew F., and Ben J. Hatchwell. 2001. "Experimental Evidence for Kin-Biased Helping in a Cooperatively Breeding Vertebrate." *Proceedings of the Royal Society B: Biological sciences* 268 (1481): 2169–74. https://doi.org/10.1098/rspb.2001.1790.

Russell, Yvan I., Josep Call, and Robin I. M. Dunbar. 2008. "Image Scoring in Great Apes." *Behavioural Processes* 78: 108–11.

Rustagi, Devesh, Stefanie Engel, and Michael Kosfeld. 2010. "Conditional Cooperation and Costly Monitoring Explain Success in Forest Commons Management." *Science* 330 (6006): 961–65. https://doi.org/10.1126/science.1193649.

Sachs, Joel L., and Ellen L. Simms. 2006. "Pathways to Mutualism Breakdown." *Trends in Ecology & Evolution* 21 (10): 585–92. https://doi.org/10.1016/j.tree.2006.06.018.

Sala, Osvaldo E., Detlef van Vuuren, Henrique M. Pereira, David Lodge, Jacqueline Alder, Graeme Cumming, Andrew Dobson, Volkmar Wolters, and Marguerite A. Xenopoulos. 2005. "Biodiversity Across Scenarios." In *Ecosystems and Human Well-Being: Scenarios, Volume 2*, edited by Stephen R. schCarpenter, Prabhu L. Pingali, Elena M. Bennett, and Monika B. Zurek, 375–440. Washington: Island Press.

Salk, Carl, Ulrich J. Frey, and Hannes Rusch. 2014. "Comparing Forests Across Climates and Biomes: Qualitative Assessments, Reference Forests and Regional Intercomparisons." *PloS one* 9 (4): e94800. https://doi.org/10.1371/journal.pone.0094800.

Sandström, Camilla, and Camilla Widmark. 2007. "Stakeholders' Perceptions of Consultations as Tools for Co-Management - a Case Study of the Forestry and Reindeer Herding Sectors in Northern Sweden." *Forest Policy and Economics* 10 (1-2): 25–35.

Sarle, Warren S. 1997. "Neural Network FAQ, Part 1 of 7: Introduction, Periodic Posting to the Usenet Newsgroup Comp.Ai.Neural-Nets." Accessed July 08, 2012. ftp://ftp.sas.com/pub/neural/FAQ.html.

Scheberle, Denise. 2000. "Moving toward Community-Based Environmental Management: Wetland Protection in Door County." *American Behavioral Scientist* 44 (4): 565–79.

Schino, Gabrielle. 2007. "Grooming and Agonistic Support: A Meta-Analysis of Primate Reciprocal Altruism." *Behavioral Ecology* 18: 115–20.

Schlager, Edella, William Blomquist, and Shui Y. Tang. 1994. "Mobile Flows, Storage, and Self-Organized Institutions for Governing Common-Pool Resources." *Land Economics* 70 (3): 294–317.

Schlager, Edella, and Elinor Ostrom. 1992. "Property-Rights Regimes and Natural Resources: A Conceptual Analysis." *Land Economics* 68 (3): 249–62.

Schlüter, Achim. 2006. "Constraints on Institutional Change in smallscale Forestry: A New Institutional Economics Perspective." *Schweizerische Zeitschrift fur Forstwesen* 157 (3-4): 84–90. https://doi.org/10.3188/szf.2006.0084.

Schlüter, Achim. 2007. "Institutional Change in the Forestry Sector—The Explanatory Potential of New Institutional Economics." *Forest Policy and Economics* 9 (8): 1090–99. https://doi.org/10.1016/j.forpol.2006.11.001.

Schlüter, Achim, and Róger Madrigal. 2012. "The SES Framework in a Marine Setting: Methodological Lessons." *Rationality, Markets and Morals* 3: 158–79. http://hdl.handle.net/10535/8584.

Schurr, Chistoph. 2006. *Zwischen Allmende und Anti-Allmende.* Freiburg: Dissertation.

Seibold, Sebastian, Martin M. Gossner, Nadja K. Simons, Nico Blüthgen, Jörg Müller, Didem Ambarlı, Christian Ammer et al. 2019. "Arthropod Decline in Grasslands and Forests Is Associated with Landscape-Level Drivers." *Nature* 574 (7780): 671–74. https://doi.org/10.1038/s41586-019-1684-3.

Semmann, Dirk, Hans-Jürgen Krambeck, and Manfred Milinski. 2003. "Volunteering Leads to Rock-Paper-Scissors Dynamics in a Public Goods Game." *Nature* 425 (6956): 390–92. https://doi.org/10.1038/nature01986.

Shiferaw, Bekkele, Tewodros A. Kebede, and Ratna V. Reddy. 2008. "Community Watershed Management in Semi-Arid India: The State of Collective Action and Its Effects on Natural Resources and Rural Livelihoods." Accessed July 08, 2012.

Shivakoti, Ganesh P., and Elinor Ostrom, eds. 2002. *Improving Irrigation Governance and Management in Nepal.* San Francisco: Institute for Contemporary Studies.

Siebert, Stefan, Burke Jacob, Faures Jean-Marc, Frenken Karen, Hoogeveen Jippe, Döll Petra, and Portmann Felix. 2010. "Groundwater Use for Irrigation—a Global Inventory." *Hydrology and Earth System Sciences* 14 (10): 1863–80. https://doi.org/10.5194/hess-14-1863-2010.

Soliveres, Santiago, Fons van der Plas, Peter Manning, Daniel Prati, Martin M. Gossner, Swen C. Renner, Fabian Alt et al. 2016. "Biodiversity at Multiple Trophic Levels Is Needed for Ecosystem Multifunctionality." *Nature* 536 (7617): 456–59. https://doi.org/10.1038/nature19092.

Sosis, Richard, and Bradley J. Ruffle. 2004. "Ideology, Religion, and the Evolution of Cooperation: Field Experiments on Israeli Kibbutzim." *Research in Economic Anthropology* 23: 89–117.

Stevenson, Todd C., and Brian N. Tissot. 2014. "Current Trends in the Analysis of Co-Management Arrangements in Coral Reef Ecosystems: A Social-Ecological Systems

Perspective." *Current Opinion in Environmental Sustainability* 7: 134–39. https://doi. org/10.1016/j.cosust.2014.02.002.

Stoker, Gerry. 1998. "Governance as Theory: Five Propositions." *International Social Science Journal* 50 (155): 17–28. https://doi.org/10.1111/1468-2451.00106.

Stoop, Jan, Charles N. Noussair, and Daan van Soest. 2009. "From the Lab to the Field: Public Good Provision with Fishermen." Accessed April 01, 2020. http://hdl. handle.net/10535/1645.

Sunderlin, William D., Arild Angelsen, Brian Belcher, Paul Burgers, Robert Nasi, Levania Santoso, and Sven Wunder. 2005. "Livelihoods, Forests, and Conservation in Developing Countries: An Overview." *World Development* 33 (9): 1383–1402. https:// doi.org/10.1016/j.worlddev.2004.10.004.

Sutter, Matthias, Stefan Haigner, and Martin G. Kocher. 2008. "Choosing the Carrot or the Stick?—Endogenous Institutional Choice in Social Dilemma Situations." Accessed July 08, 2012. http://editorialexpress.com/cgi-bin/conference/download.cgi?db_ name=res2006&paper_id=395.

Tang, Shui Y. 1989. "Institutions and Collective Action in Irrigation Systems." Accessed September 30, 2011. http://dlc.dlib.indiana.edu/dlc/handle/10535/3596.

Tang, Shui Y. 1991. "Institutional Arrangements and the Management of Common-Pool Resources." *Public Administration Review* 51 (1): 42–51.

Tang, Shui Y. 1992. *Institutions and Collective Actions: Self-Governance in Irrigation.* San Francisco: Institute for Contemporary Studies.

Taylor, Paul W. 1986. *Respect for Nature: A Theory of Environmental Ethics.* Princeton: Princeton University Press.

Thiel, Andreas, Muluken E. Adamseged, and Carmen Baake. 2015. "Evaluating an Instrument for Institutional Crafting: How Ostrom's Social-Ecological Systems Framework Is Applied." *Environmental Science & Policy* 53: 152–64. https://doi.org/ 10.1016/j.envsci.2015.04.020.

Thomson, James T., David Feeny, and Ronald J. Oakerson. 1992. "Institutional Dynamics: The Evolution and Dissolution of Common-Property Resource Management." In Bromley, Feeny, Peters, Gilles, Oakerson, Runge, and Thomson, eds. 1992, 129–60.

Thrush, Simon F., Giovanni Coco, and Judi E. Hewitt. 2008. "Complex Positive Connections Between Functional Groups Are Revealed by Neural Network Analysis of Ecological Time Series." *American Naturalist* 171 (5): 669–77. https://doi.org/10.1086/ 587069.

Tinbergen, Nico. 1963. "On Aims and Methods of Ethology." *Zeitschrift für Tierpsychologie* 20: 410–33.

Torres-Guevara, Luz E., and Achim Schlüter. 2016. "External Validity of Artefactual Field Experiments: A Study on Cooperation, Impatience and Sustainability in an Artisanal Fishery in Colombia." *Ecological Economics* 128: 187–201. https://doi.org/10.1016/ j.ecolecon.2016.04.022.

Trivers, Robert L. 1971. "The Evolution of Reciprocal Altruism." *Quarterly Review of Biology* 46 (1): 35–57.

Trivers, Robert L. 1985. *Social Evolution.* Menlo Park, California: Benjamin/Cummings.

Tucker, Catherine M. 2010. "Learning on Governance in Forest Ecosystems: Lessons from Recent Research." *International Journal of the Commons* 4 (2): 687–706.

Tucker, Catherine M., James C. Randolph, and E J. Castellanos. 2007. "Institutions, Biophysical Factors and History: An Integrative Analysis of Private and Common Property Forests in Guatemala and Honduras." *Human Ecology* 35: 259–74.

Tucker, Catherine M., James C. Randolph, Tom Evans, Krister P. Andersson, Lauren Persha, and Glen M. Green. 2008. "An Approach to Assess Relative Degradation in Dissimilar Forests: toward a Comparative Assessment of Institutional Outcomes." *Ecology and Society* 13 (1): 1–21.

Uhl, Matthias, and Eckart Voland. 2002. *Angeber haben mehr vom Leben.* Heidelberg: Spektrum, Akademischer Verlag.

van Laerhoven, Frank. 2010. "Governing Community Forests and the Challenge of Solving Two-Level Collective Action Dilemmas—A Large-N Perspective." *Global Environmental Change* 20 (3): 539–46. https://doi.org/10.1016/j.gloenvcha.2010.04.005.

Varughese, George, and Elinor Ostrom. 2001. "The Contested Role of Heterogeneity in Collective Action: Some Evidence from Community Forestry in Nepal." *World Development* 29 (5): 747–65.

Velicer, Gregory J., Lee Kroos, and Richard E. Lenski. 2000. "Developmental Cheating in the Social Bacterium Myxococcus Xanthus." *Nature* 404: 598–601.

Vogt, Jessica M., Graham B. Epstein, Sarah K. Mincey, Burnell C. Fischer, and Paul McCord. 2015. "Putting the 'E' in SES: Unpacking the Ecology in the Ostrom Social-Ecological System Framework." *Ecology and Society* 20 (1): 55–65. https://doi.org/10.5751/ES-07239-200155.

Voland, Eckart. 2003. "Eigennutz und Solidarität—das konstruktive Potenzial biologisch evolvierter Kooperationsstrategien im Globalisierungsprozess." *Zeitschrift für internationale Bildungsforschung und Entwicklungspädagogik* 26 (4): 15–20.

Voland, Eckart. 2013. *Soziobiologie—Die Evolution von Kooperation und Konkurrenz.* Heidelberg: Springer.

Voland, Eckart, and Wulf Schiefenhövel, eds. 2009. *The Biological Evolution of Religious Mind and Behavior.* Heidelberg: Springer.

Vollan, Björn. 2008. "Socio-Ecological Explanations for Crowding-Out Effects from Economic Field Experiments in Southern Africa." *Ecological Economics* 67: 560–73. https://doi.org/10.1016/j.ecolecon.2008.01.015.

Vörösmarty, Charles J., Pamela Green, Joseph Salisbury, and Richard B. Lammers. 2000. "Global Water Resources: Vulnerability from Climate Change and Population Growth." *Science* 289 (5477): 284–88. https://doi.org/10.1126/science.289.5477.284.

Wade, Robert. 1992. "Common-Property Resource Management in South Indian Villages." In Bromley, Feeny, Peters, Gilles, Oakerson, Runge, and Thomson, eds. 1992, 207–28.

Wade, Robert. 1994. *Village Republics: Economic Conditions for Collective Action in South India.* San Francisco: Institute for Contemporary Studies.

Walker, Brian, C S. Holling, Stephen R. Carpenter, and Ann Kinzig. 2004. "Resilience, Adaptability and Transformability in Social–ecological Systems." *Ecology and Society* 9 (2): 5. http://www.ecologyandsociety.org/vol9/iss2/art5.

Walker, James M., Roy Gardner, Andrew Herr, and Elinor Ostrom. 2000. "Collective Choice in the Commons: Experimental Results on Proposed Allocation Rules and Votes." *Economic Journal* 110: 212–34.

Waylen, Kerry A., Anke Fischer, Philip K. McGowan, Simon J. Thirgood, and E J. Milner-Gulland. 2010. "Effect of Local Cultural Context on the Success of Community-Based Conservation Interventions." *Conservation Biology* 24 (4): 1119–29.

Wedekind, Claus, and Manfred Milinski. 2000. "Cooperation through Image Scoring in Humans." *Science* 288 (5467): 850–52.

Wenseleers, Tom, A Helanterä, A Hart, and Francis L. W. Ratnieks. 2004. "Worker Reproduction and Policing in Insect Societies: An ESS Analysis." *Journal of Evolutionary Biology* 17: 1035–47.

Werthmann, Christine, Anne Weingart, and M Kirk. 2008. "Common-Pool Resources— a Challenge for Local Governance Experimental Research in Eight Villages in the Mekong Delta of Cambodia and Vietnam." Accessed April 01, 2020. https://www.academia.edu/download/48447441/Common-Pool_Resources_-_A_Challenge_For_20160830-3065-17owucr.pdf.

West, Stuart A., Claire El Mouden, and Andy Gardner. 2011. "Sixteen Common Misconceptions about the Evolution of Cooperation in Humans." *Evolution and Human Behavior* 32 (4): 231–62. https://doi.org/10.1016/j.evolhumbehav.2010.08.001.

West, Stuart A., Ashleigh S. Griffin, and Andy Gardner. 2007. "Social Semantics: Altruism, Cooperation, Mutualism, Strong Reciprocity and Group Selection." *Journal of Evolutionary Biology* 20 (2): 415–32. https://doi.org/10.1111/j.1420-9101.2006.01258.x.

White, Andy, and Alejandra Martin. 2002. *Who Owns the World's Forests? Forest Tenure and Public Forests in Transition.* Washington: Forest Trends.

Widrow, Bernard, David E. Rumelhart, and Michael E. Lehr. 1994. "Neural Networks: Applications in Industry, Business and Science." *Communications of the ACM* 37 (3): 93–105.

Wiessner, Polly. 2009. "Experimental Games and Games of Life among the Ju/'hoan Bushmen." *Current Anthropology* 50 (1): 133–38. https://doi.org/10.1086/595622.

Willems, Erik P., Barbara Hellriegel, and Carel P. van Schaik. 2013. "The Collective Action Problem in Primate Territory Economics." *Proceedings of the Royal Society B: Biological sciences* 280 (1759): 20130081. https://doi.org/10.1098/rspb.2013.0081.

Williams, Graham. 2011. *Data Mining with Rattle and R: The Art of Excavating Data for Knowledge Discovery.* Heidelberg: Springer.

Williamson, Oliver E. 1975. *Markets and Hierarchies: Analysis and Antitrust Implications: A Study in the Economics of Internal Organization.* New York: Free Press.

Williamson, Oliver E. 2000. "The New Institutional Economics: Taking Stock, Looking Ahead." *Journal of Economic Literature* 38 (3): 595–613.

Wilson, David S., David Near, and Ralph R. Miller. 1996. "Machiavellianism: A Synthesis of the Evolutionary and Psychological Literatures." *Psychological Bulletin* 119 (2): 285–99.

Wilson, Edward O. 2013. *The Social Conquest of Earth.* New York, London: Liveright.

Wit, Arjan p., and Norbert L. Kerr. 2002. "'Me versus Just Us versus Us All' Categorization and Cooperation in Nested Social Dilemmas." *Journal of Personality and Social Psychology* 83 (3): 616–37.

Wollenberg, Eva K., Leticia Merino, Arun Agrawal, and Elinor Ostrom. 2007. "Fourteen Years of Monitoring Community-Managed Forests: Learning from IFRI's Experience." *International Forest Review* 9 (2): 670–84.

Xu, Li, Dora Marinova, and Xiumei Guo. 2015. "Resilience Thinking: A Renewed System Approach for Sustainability Science." *Sustainability Science* 10 (1): 123–38. https://doi.org/10.1007/s11625-014-0274-4.

Yamagishi, Toshio, Yutaka Horita, Haruto Takagishi, Mizuho Shinada, Shigehito Tanida, and Karin S. Cook. 2009. "The Private Rejection of Unfair Offers and Emotional Commitment." *Proceedings of the National Academy of Sciences of the United States of America* 106 (28): 11520–23.

Yamagishi, Toshio, Shigeru Terai, Toko Kiyonari, Nobuhiro Mifune, and Satoshi Kanazawa. 2007. "The Social Exchange Heuristic: Managing Errors in Social Exchange." *Rationality and Society* 19 (3): 259–91.

Yeh, I-Cheng, and Wei-Lun Cheng. 2010. "First and Second Order Sensitivity Analysis of MLP." *Neurocomputing* 73 (10-12): 2225–33. https://doi.org/10.1016/j.neucom.2010.01.011.

Young, Andrew J., Anne A. Carlson, Steven L. Monfort, Andrew F. Russell, Nigel C. Bennett, and Tim H. Clutton-Brock. 2006. "Stress and the Suppression of Subordinate Reproduction in Cooperatively Breeding Meerkats." *Proceedings of the National Academy of Sciences of the United States of America* 103 (32): 12005–10. https://doi.org/10.1073/pnas.0510038103.

Zahavi, Amotz, and Avishag Zahavi. 1997. *The Handicap Principle: A Missing Piece of Darwin's Puzzle.* Oxford: Oxford University Press.

Zefferman, Matthew R. 2014. "Direct Reciprocity Under Uncertainty Does Not Explain One-Shot Cooperation, but Demonstrates the Benefits of a Norm Psychology." *Evolution and Human Behavior* 35 (5): 358–67. https://doi.org/10.1016/j.evolhumbehav.2014.04.003.